SpringerBriefs in Latin American Studies

More information about this series at http://www.springer.com/series/14332

Guillaume Leturcq

Dams in Brazil

Social and Demographical Impacts

Springer

Guillaume Leturcq
Campinas, São Paulo, Brazil

ISSN 2366-763X ISSN 2366-7648 (electronic)
SpringerBriefs in Latin American Studies
ISBN 978-3-319-94627-6 ISBN 978-3-319-94628-3 (eBook)
https://doi.org/10.1007/978-3-319-94628-3

Library of Congress Control Number: 2018946587

Printed on acid-free paper

This Springer imprint is published by the registered company Springer Nature Switzerland AG
The registered company address is: Gewerbestrasse 11, 6330 Cham, Switzerland

To Teresa and Margot. Je vous aime.

Foreword

The book by Dr. Guillaume Leturcq, on Dams in Brazil, is an important contribution. It is the first book to discuss the development of hydroelectricity in Brazil and to focus on both the south of Brazil and the north. It provides a rich historical review of these developments and ties these developments to the geopolitical ambitions of Brazil's military and civilian leaders since the 1940s, and unquestioned since then by its presidents and leadership. Even the Workers' Party, led by Lula, did not question these priorities during their 12 years in office. The most emblematic dams discussed by the author were started during the military dictatorship and were part of the Giant Brazil plans for rapid economic development and accelerated industrialization (e.g., Tucurui, Itaipu, Sobradinho), and in a period of military rule, rights of people in the way of dam construction could be overlooked with impunity. What is truly surprising is the repetition of these behaviors under the leadership of the Workers' Party, which forced the approval of Belo Monte over the 20-year opposition of social movements, indigenous groups, scientists, and environmentalists—and which led to egregious lack of consultation, social, and environmental impacts as had not been seen before, and lack of compensation for many of the people negatively impacted by the dam.

From the start, hydropower in Brazil has been tied to either addressing the problem of droughts in Northeast Brazil, or the power needs of cities and industry in Minas Gerais, Rio de Janeiro, and Sao Paulo. Over time, other large cities along the coast of Brazil (Fortaleza, Recife, Salvador, Vitoria, Curitiba, Porto Alegre) also came to demand more energy as they grew in population and economic activity. The location of most of the 200+ dams in Brazil is in the southeast and that makes sense since that is where the largest demand is, and this avoids loss of energy in transmission. That potential is largely tapped already. Beginning with Tucurui, the government sets its sights on developing the huge hydropower potential in the rivers of Amazonia.

The future of dams in Brazil seems poised to be increasingly influenced by the growing role of Chinese firms in this sector. They are already the second largest actors in this sector and growing quickly in both the production and transmission of energy in Brazil. This ties Brazil to the global economy even more than already and

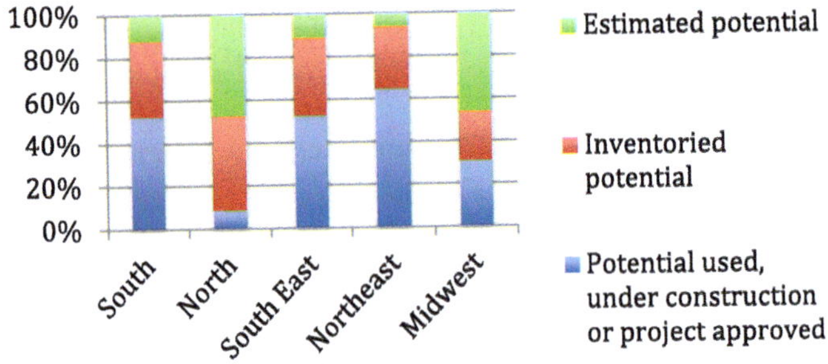

Fig. 1 Evolution of the share of hydroelectricity in Brazil's installed electrical capacity, 2006. *Data source* Ministry of Mines and Energy (MME) and IBGE

Fig. 2 Map of location of the 219 hydroelectric plants in Brazil. *Data source* ANEEL 2017

of the country, from the north to the south (*Cf.* Fig. 2). The location of the dams is a key element in the analysis of the effects on the populations and the Brazilian society as a whole, and therefore a key criterion in this book.

The large number of dams and the reservoirs that accompany them mark the country. The focus of this book is to discover exactly at what levels and how intensely the effects are felt by the population, the society, and the territory. We shall study in detail the effects of dams on the Brazilian territory, by looking at various scales, time periods, and impact intensities, in order to better understand how dams continue to mark the Brazilian society so deeply and with such indelible imprints.

The construction of dams necessarily leaves a mark on the territories and Brazil as a whole. We will first address the subject from a historical, geographical, and landscape point of view. The geographical study of the territory and space helps to better understand the dimension and the influence of the structures. The second part of the book will focus on the social and human issues to better understand the impacts on the populations. The topics of migration, living space, adaptation, and identity have been chosen to explain the social marks made by dams. Finally, the last part of the book will provide a north–south territorial comparison of the effects of dams on very different geographical spaces, but with clear similarities and differences.

electricity. Major technical developments and industrialization are the main reasons for the rapid multiplication of hydroelectric dams.

The World Commission on Dams (WCD) defines a large dam as a structure with a minimum height of 5 m and with a reservoir volume of more than 3 million cubic meters (WCD 2000). This refines a definition of The International Commission on Large Dams (ICOLD) which, in 1928, described large dams as structures higher than fifteen meters. Following on from the WCD definition, the new small dam category can be described as structures lower than 5 m and/or with a reservoir volume under 3 million cubic meters.

For the Brazilian energy sector, the definition of large dams is based on the amount of energy produced. According to the Atlas of Energia Elétrica do Brasil, the power of production determines the size of the hydroelectric dam: If a plant produces more than 30 megawatts (MW), it is considered large; if it produces less than 30 MW, it is classified as a small hydroelectric dam or a hydropower generating plant for installations with an output below 1 MW (ANEEL 2008). Here, we would like to make it clear that there is a difference between the installed capacity and the actual generation of energy: Installed capacity is the amount of energy that would be generated in a year if all the turbines were to run at full capacity during 365 days.

These three categories of electrical installations coexist throughout Brazil, each of which is accompanied by a dam used to retain the watercourse in order to supply the turbines with water. Since our focus in this work is only on large dams, we will use the terms hydroelectric plant and dam interchangeably to speak of one and the same type of structure.

The construction of the dam, usually in concrete, aims to control the watercourse. For a long time, Brazil built hydroelectric plants with dams creating a reservoir of water much greater than the natural river bed. These structures had a significant impact on the people and the environment. Due to pressures from actors outside the field of hydroelectricity, the companies responsible for the construction of dams decided to move toward "run-of-river" dams, which use the original bed of the river and a slight lake of accumulation, in no way comparable to the dams previously built. From an engineering point of view, run-of-river dams are still controversial: They do not always store enough water, making it difficult to manage extreme climatic periods (rain and drought periods), sometimes requiring the construction of additional retention dams upstream.

The first dams built in Brazil were not aimed at producing electricity but at combatting drought. As early as the nineteenth century, the authorities in power became aware that repeated droughts in the Nordeste region were hindering development. Consequently, the National Department of Works against Drought decided to analyze the situation and made some recommendations (construction of a reservoir and diversion of the San Francisco River), which barely got any response. The great drought that took place between 1877 and 1889 greatly influenced the construction of the first dam in Brazil, the Cedro Dam (state of Ceara), which began in 1884 and ended in 1906 (Mello 2011).

The first hydroelectric plant in Brazil was built at the end of the nineteenth century in the state of Minas Gerais (1883). Backed by a small dam, creating about 5 m of head, the plant, located on the Jequitinhonha River, is associated with the Mines and Metallurgy School de Ouro Preto. This first development represents the starting point of a movement to build other plants to support the mining activities of the region. A few years later (1889), a hydroelectric plant (Marmelos) was used for the first time for public illumination in Juiz de Fora (Minas Gerais). Thus began the development of this technology for public and private purposes (the location of the mentioned dams can be observed in Fig. 1.1).

The end of the nineteenth and the beginning of the twentieth century marked a period of creation and development of urban centers and of rapid industrialization. The demand for energy was increasing, and electricity generation moved from an experimental phase to a production phase. Thus began what is still today called the Brazilian electricity sector. We are using this term in our work to better explain how this industrial branch is organized. Production increased rapidly at the turn of the twentieth century, from 5 MW in 1900 to more than 30 MW (data from IPEA 2014, http://www.ipeadata.gov.br) the following year. Political and legislative

Fig. 1.1 Location of the mentioned dams. *Data source* ANEEL (2017)

conditions were put into place. The Brazilian Constitution of 1891 granted municipalities the right to distribute concessions for the use of waterfalls and the production of electricity. Later, a law (1903), followed by a decree (1904), gave the federal government the authority to use the hydraulic energy of the rivers for public service.

The country was booming but did not yet have enough investors, so foreign companies were invited to invest, notably by establishing themselves in the urban centers of São Paulo and Rio de Janeiro. In parallel, Brazilian private and public companies were also created to offer a variety of services, such as transportation, street lighting, electricity production and distribution. The Light Company, originally from Canada, was the first one to really put its mark on the sector through its influence and investments. It set itself apart by the diversity of its activities—in the production of electricity, with the construction of dams, in the distribution of electricity in major urban centers, and also for public transport, with trams in Rio de Janeiro.

With the development of the sector, the Brazilian state began to observe and control, initially creating, within the Ministry of Agriculture, the Federal Commission of Hydraulic Forces (1920), before writing the Water Code in 1934. This was the first regulation of the energy sector which declared the state sole owner of the territorial waters and therefore required authorization to exploit resources. At the same time, energy production was constantly increasing to meet demand, doubling the production capacity each decade until the 1930s (IPEA 2014). A few years later, in 1939, the National Council of Water and Electrical Energy was formed. Its role was to manage the exploitation and use of energy in Brazil and to implement the provisions included in the Water Code. The idea was to strengthen the role and position of the state in this sector, which previously belonged to private companies. This corresponded to the political will to centralize power with President Getúlio Vargas (1937–1945) and his establishment of the "Estado Novo." The instigation of a concession regime was the hallmark of the Brazilian model for the exploitation of natural resources. There was a real planning vision for the future, both for the electricity sector and for the country and its growing energy needs. At this time, there was a growing awareness of the hydroelectric potential of the country because of its characteristics in terms of relief, climate, and geomorphology (Queiroz 2013). The future of the country in terms of energy was decided, and it depended on hydroelectricity. Since the beginning of the twentieth century until 1960, the installed capacity of hydroelectricity accounted for at least 80% of the country's total electricity capacity (IPEA 2014).

Wishing to confirm its new position on a regional level, the federal state created in 1945 the CHESF, São Francisco's Hydroelectric Company, whose aim was to operate the watershed of the São Francisco River in the Nordeste region of Brazil. In 1948, the CHESF began construction of the first large hydroelectric plant in the region, Paulo Afonso I, which started operating in 1954 (Fundação Getulio Vargas 2009). The hydroelectric potential of the river São Francisco, coveted for several years, began to be harnessed over the following decades. From the 1950s, the second Vargas government chose to invest in hydroelectricity to boost power generation and provided general guidance. In 1954, a Federal Electrification Fund was set up, financed by a single tax on electricity based on consumption, thus becoming the main source of income for investments in the sector. In the same year, a new National Electrification Plan was presented to the Congress. It confirmed the state's investment in the electricity sector (Corrêa 2003) and proposed the creation of the state enterprise Eletrobras. Congress opposed this creation, and it took another 8 years for Eletrobras to be officially established.

During the 1950s, there was much debate on the future of the sector, particularly at the political level, and engineers played a fundamental role in it, as described by Mari Letícia Corrêa in her thesis. No substantial progress was made during this decade, and the electricity generating fleet grew as a result of federal government interventions. The second half of the 1950s saw the development of the institutional structure, with mainly the creation of state enterprises in the federal states (Corrêa 2003). This decade saw the consumption of electricity increase as a result of population growth, increased urbanization, and the military government's economic policy.

1.1.2 Energy and Politics

In the early 1960s, the sector continued its structuring in which two institutions played a very important role. In 1960, the Ministry of Mines and Energy (MME) was established and given powers previously belonging to the Ministry of Agriculture.

In 1962, Eletrobras was finally created and linked to the MME. The company holds a prominent place in the Brazilian electricity sector with regard to the production, transmission, and even distribution of electricity. It coordinates the "technical, administrative, and financial aspects of the country's electricity sector" (Martins 2008). Eletrobras is a holding company which manages the states of the federation's public enterprises. It is an essential player in Brazil's future energy policies.

The 1960s were marked by a profound change in the national policy, which upset all economic sectors, including, of course, the electricity sector. The junta, following a "development" strategy, undertook major projects such as the construction of hydroelectric dams. In this logic, two National Development Plans were established in 1967 and 1976, in which dams played a prevalent role. The Brazilian government opened up to foreign investors for these projects. This decade saw the consumption of electricity increase.

Brazil took part in an international wave of "dam ideology," an economic rationale for the construction of major structures, carried out by international organizations (McCully 2001), such as the World Bank or the International Monetary Fund (FMI). Those projects were politically endorsed by local elites, especially the military in power in the case of Brazil (Locatelli 2014). Because of the coup, different political orientations, project development issues, and the quest for funding, the major projects planned by the military were delayed until the second half of the 1970s (Itaipu or Sobradinho for example).

Social and environmental issues only really started to emerge in the early 1980s, toward the end of the military rule. The 1980s also marked a new period of change in the electricity sector, with increased concerns for environmental and social issues.

In the early 1980s, the country experienced a deep political, financial, and economic crisis. The whole structure of the country changed with the return of democracy, and a phase of transition began in which certain rules of the economic game were rewritten. The international context also started to weigh more heavily, with new environmental concerns emerging. Ongoing projects were completed by dealing with the conflicts arising in the best possible way. No new dam construction was undertaken and national electricity production capacity remained steady, while consumption continued to rise. A new electricity rationalization program was put in place, similar to that of the 1950s.

With the transition, a whole new political, institutional, and legislative framework was established, which lead to the constitution of 1988. A new institutional framework for the country's electricity activities was set up. Demand also arose for assessing the environmental impacts of development projects.

The first stage, still in a period of military rule, was the creation, in 1981, of the National Environmental Policy, which for the first time established a system of environmental licenses for potentially polluting activities (Mello 2011). However, it was not until 1986 and a publication in the official gazette that the measure came into force. In 1981, the National Environment System (SISNAMA) was also created. It brought together all environmental organizations at the federal, state, and municipal levels. Finally, in 1981, the National Council for the Environment (CONAMA) was also established to look after the regulation and legislation of the PNMA. In 1985, the first Ministry of the Environment was constituted (MMA). The first environmental institutions appeared at the end of the military regime and the return to democracy accelerated the movement. After 1985, the whole landscape of the Brazilian electricity sector evolved, following other changes in the country.

As a continuation of the preliminary environmental initiatives of the early 1980s, CONAMA published its first resolution in 1986 defining what an environmental impact is. This new resolution, which is still applicable today, requires two preliminary studies to be carried out before starting work on a potentially polluting project. The objectives of these studies are to evaluate the likely impacts both of the construction site and of the final structure. The first study is called the Environmental Impact Assessment (EIA), and the second, the Environmental Impact Report (RIMA), is a summary of the first. They provide, according to the CONAMA resolution, an environmental diagnosis of the building site and an assessment of the likely post-construction environmental consequences. The studies must be accompanied by an Environmental Basic Project (PBA) including plans to mitigate negative effects on the environment and a support plan for dealing with all types of impacts (positive/negative, direct/indirect, short/medium/long term) (Martins 2008).

The same year (1986), the CONAMA released several resolutions describing in detail the steps required for the application of environmental licenses (LA). Their implementation completely changed the planning for hydroelectric structures. To obtain an environmental license, three steps are required to be granted three preliminary licenses:

- LP—Prerequisite License: Sought at the end of a project site's study phase, it is valid for a maximum of 5 years.
- LI—Installation License: This license validates the project and allows construction to start. It must be preceded by the presentation of the PBA. It has a maximum validity of 6 years.
- LO—Operation License: Once construction is complete, the LO allows commencement of the activities. It may be granted for a period of 4 to 10 years and may be renewed.

Licenses are granted by environmental monitoring bodies, at the state or federal level, for example, in the case of dams built on a river forming the border between two states. The Brazilian Institute of Environment and Renewable Natural Resources (IBAMA) is responsible for environmental licensing at the federal level, while the Secretariat for the Environment deals with it at the state level.

The environmental licensing system has been questioned for various reasons. The first concerns the non-compliance, by private companies, with the law and advocated standards, in particular, due to the lack of inspection and support during construction work (Moreira 2006). For C. Caubet, Professor of Law, there has not been a single case of full compliance with the environmental law for the construction of hydroelectric dams in Brazil (Caubet 2004). The environmental law is only really a framework, which does not fundamentally prevent the construction of structures. According to C. Caubet, the laws are really only "apparent standards," which means they are not actually enforced and basically ineffective. He gives the example of the Barra Grande Dam, for which the justice system recognized that licenses were irregularly obtained, but only after the dam had been completed. The construction of a dam is never questioned during the process of obtaining licenses. Environmental studies allow a better understanding of the negative and positive effects of the dams but do not influence their construction. The WCD (2000) states that impact studies must be carried out to "offset or mitigate the expected impacts and make them acceptable when the decision to build has already been made" (International Network Rivers 2008). Numerous dam construction examples demonstrate the fragility of the laws and their applications (Belo Monte or Jirau, among others), especially in the face of economic and political interests. Environmental laws are benchmarks that provide guidance but whose principles and application are sometimes flouted.

The first legislative measures taking into account the environment greatly influenced national policies and set the trend for the new constitution of 1988: All water resources (lakes, rivers, aquifers, springs, marine waters, etc.) are thus defined as property of the state which has therefore the power to use, directly or indirectly, the hydraulic potential of these resources.

The electricity sector adapted to the changes, and in 1986, Eletrobras published a master plan (PDMA) for the protection and improvement of the environment for electricity projects and services. This first environmental policy for the sector was based on four key aspects: environmental sustainability, regional integration, coordination between institutions and with the company, and management efficiency (Mello 2011). The PDMA is supported by an Environmental Advisory Committee within Eletrobras (Consultative Committee of Meio Ambiente). The electricity sector was therefore responsive to the changes underway and preparing for the coming decades. At the beginning of the twenty-first century, environmental laws continued to evolve but with no major changes. It is even possible to identify the beginnings of a backtracking in the sphere o environmental laws. Taking advantage of the chaotic political situation, some representatives, driven by different lobbies and pressure groups (such as landowners and contractors), are trying to shorten the process of environmental licensing (draft bill 65), simplifying it as much as possible (Fearnside 2016b). Opposition groups started to organize to prevent those measures, but threats are still present.

An ecological awareness was starting to develop regarding water resources, not only in the electricity sector but also more globally within society and among policy makers. A range of measures was taken to improve management of these resources. Law N° 9.433 of 1997 established a National Council of Hydric Resources (CNRH) whose responsibility was to develop a national water resources policy. A National Plan of Hydric Resources (PNRH) was established under the same law. The PNRH defined "a set of general guidelines and objectives for the management and planning of the use of water resources" (Martins 2008). It required the states of the federation to draw up plans for water resources in the states' catchments. Relying on the social participation of citizens (Wolkmer and Pimmel 2013), its governance is considered democratic, at least in theory. The 1997 Act recognized water as a "limited natural resource" and a vulnerable resource and gave it an economic value, therefore encouraging a multiplicity of uses. In addition, the National Water Resources Management System (SINGREH) was created in order to collect and disseminate information on existing resources. The 1997 Act also required a license application for companies wishing to exploit water resources. After 9 years of discussions and negotiations, the PNRH was finally approved by the CNRH in January 2006.

In 2000, the government founded the National Agency for Water (ANA), a government agency responsible for the monitoring, control, and evaluation of activities related to water resources. The agency is also responsible for the implementation of the National Water Resources Plan (PNRH).

All developments from the end of the 1980s were a prelude to the privatization of the electricity sector, which began in 1992 and slowly took shape during the decade. The first step was the National Plan of Privatization (PND) which started by selling the states' public companies responsible for the distribution of energy (Gomes et al. 2006). Legal provisions made in 1993 and 1994 gave the National Congress the power to authorize sales of businesses. In 1995, a national privatization council was created by the government of President Fernando Henrique Cardoso. The first sales were made to foreign companies, mainly American.

The next step began with the preparation of the sale of power generation companies. To this end, a control agency was created: the National Agency for Electrical Energy (ANEEL). Attached to the MME, it is the last in a series of agencies that came into existence about a century ago: the Water Directorate (1933), created with the Water Code; the Water Department, which later became the Water Division, still within the Ministry of Agriculture; and the National Council of Water and Electrical Energy (1939) which in 1965 became the National Department of Water and Energy. ANEEL's aim is to monitor and order the production, transmission, and marketing of energy. The three main phases of the energy market therefore came under the control of ANEEL which is also responsible for monitoring the market and its players.

The privatization phase continued with the use of an English firm (Coopers & Lybrand) to manage the privatization of the production sector. The process of selling the production companies, controlled by the state, began in 1997, and very rapidly Eletrobras was fully reorganized, including all the state companies under its control. Eletrobras also ceased to be in charge of energy transmission which became the responsibility of a new management agency, the National Operator of the System (ONS).

Finally, in 2004, a reform of the Brazilian electricity sector once again allowed state-owned enterprises such as Eletrosul to produce electricity and to make its return for the construction and management of hydroelectric dam. This reform was the result of the last major production crisis in the sector. In 2000 and 2001, the crisis was so profound that a new rationing had to be put in place and the government asked Brazilians to reduce their electricity consumption by 20%. This plan was a success for the government as the country managed to meet the expectations of the central government with a greater than expected decline in consumption.

In just over a century, Brazil went through various phases in electricity generation and environmental management. The relationship between these two areas evolved over the years. Initially (before 1930), foreign companies could produce electricity without any constraint. The economic and financial crisis of the 1930s led to the second phase, with the federal government taking over the entire energy sector. Through legislative frameworks and governmental entities, the state took control of the sector for nearly 55 years (from 1930 to 1985). The last phase was the privatization and sale to local and foreign investors. The state sold its businesses and reorganized itself to try and respond to the new economic framework and environmental concerns. However, the sector may still not be stable, because the 2001 crisis was profound. According to S. Meritet, "the reform is no yet complete, particularly in terms of defining an appropriate regulatory framework, but it is fundamental for the long-term reorganization of the industry" (Meritet 2004). The political wills in 2017 target a new privatization of the sector, notably of Eletrobras.

1.1.3 Significant Dams

The history of hydroelectricity in Brazil is marked by a few significant projects, which have marked the hydroelectric landscape for various reasons, such as their location, their international implications, their role in the electricity sector. Each of these projects has influenced and modified the history of the sector in its own way. Four of them in particular provide a better understanding of how they represent and illustrate an entire industry.

Tucuruí

The Tucuruí Dam marked history as the first major hydroelectricity project in the Brazilian Amazon. Started and completed during the military dictatorship, the first construction phase took 10 years, between 1974 and 1984. A second phase began in the late 1990s to increase the installed capacity. The impacts of the second phase were in no way comparable to those of the first stage. The dam was built on the Tocantins River, in the middle of the state of Pará (Fearnside 1999), in a small municipality which counted less than 9000 inhabitants in 1970 (IBGE—Brazilian Institute of Geography and Statistics). Its main objective was to provide energy to aluminum smelters (Fearnside 2016a). This project, launched in 1982, aimed to exploit the mineral resources of an entire region and therefore required a close source of energy.

The construction of the Tucuruí Dam dates back to a political era that has now ended in Brazil and is very significant, from an environmental point of view, because of its location in the heart of the world's largest rainforest. However, no impact assessment, positive nor negative, was requested before starting work on the Tucuruí Dam. Eletrobras only carried out an impact assessment 1 year after the start of construction (Barrow 1988). The social and environmental impacts were considered secondary at that time, and every effort was taken to not slow down the progress desired by the military government.

The period and the context also explain the lack of reliable and detailed information coming from researchers but also from the project managers (Magalhães 2007). Eletronorte estimated the number of affected families at nearly 6000. Researchers Mougeot (1988) in fact believe that between 25,000 and 35,000 people have been affected by the dam. This data relates to the first phase of the project in the early 1980s. In the second phase, the project did not foresee any new migrations but ultimately around 4000 families had to change their place of residence (Magalhães 2005). Irony and misfortune persisted for certain families which were forced to move twice in 20 years.

The uniqueness of the Tucuruí Dam is due to the diversity of the affected families. According to Magalhães (2007), in 1988 about one-third of the affected families lived in urban areas and the rest in the countryside. The total population that year counted 5031 families. About one-third of the urban families chose to radically change their place of residence and to move to the countryside. To respond to the needs of the other families, two new urban areas and eleven rural housing estates were built near the dam (Magalhães 2007). There are conflicting accounts on the number of affected people, and it is very hard to arrive at an exact final figure. Many families and family groups were excluded from the official data. Therefore, we assume that the exact number of affected families has been underestimated. It is also worth noting that the official data did not include migrations that were incorporated into the total flow of people.

Some indigenous people were also affected by the construction of the Tucuruí Dam, and three ethnic groups were partially disrupted by the reservoir (WCD 1999). These native people had already suffered from the construction of the Trans-Amazonian Highway. The Parakanãs, the main group affected by the Tucuruí Dam, saw their territory cut down and their population reduced by the various forced displacements (Destaque Amazônia 1985). The original group was separated and moved several times (WCD Tucuruí, 2000). At that time, the indigenous people were not yet organized and united to defend themselves and suffered greatly from large construction projects such as the Tucuruí Dam.

Fishermen endured reduced production (WCD 1999 and Vainer 2007) and requested compensation for several years (at least 5 years). However, it was not until the second phase of the project that they started to receive fairer compensation, especially for those downstream of the dam. For the first time in Brazil, financial compensation was offered in areas downstream from a dam, notably through the Popular Plan for Sustainable Development of the Downstream Region of the Tucuruí Hydroelectric Plant (PPDJUS) (Ferreira 2008). Reservoir fishermen are now more numerous than in the 1980s and seem to have adapted by becoming professionals and integrating commercial channels, not only locally but also in other parts of the country (Cintra et al. 2013).

The Tucuruí Dam is characterized by its location in the heart of the rainforest, the first of its sort in Latin America. Despite some preliminary studies that predicted adverse effects on the environment, nothing was done to stop them (Barrow 1988). In 1977, a report entitled "Environmental Assessment of the Tucuruí Hydroelectric Project, Rio Tocantins, Amazônia," by the ecologist Robert Goodland (1977), was published but its audience was not sufficient to prevent or change the construction of the dam. The main point raised in the report was the cleaning of the vegetation in the future dam reservoir: If the vegetation does not get removed before the water level increases, negative consequences for the climate and the local environment emerge. After years of detailed research, P. Fearnside's work indicates that the accumulation of decomposing organic matter at the bottom of the reservoir results in the release of methane (2008, 2011, 2015, 2016a). The impact of methane emissions on global warming is a subject that fuels many debates, and Fearnside (2016a) states that "tropical hydroelectric power stations emit significant quantities of greenhouse gases."

Another type of pollution encountered in Lake Tucuruí came from the transformation of mercury (methylation), in the fish, which can be hazardous to health. The phenomenon has been studied in the Amazon for a long time by researcher Jean Remy Davée Guimarães who claims that the concentration of mercury in fish is moderate in dam reservoirs with the exception of Tucuruí (2001). In a detailed study of Tucuruí, Petri Porvari stated that 92% of predatory fish samples showed

precious help of the Catholic Church, notably with the Pastoral Commission of the Earth (CPT) and the trade unions, especially of rural workers. The change of residence was particularly disastrous for the families, especially the first 5 days. During the journey to their new homes, either by boat or by bus, the affected families lost many belongings and several animals, for which no compensation was ever paid, and therefore, they had to begin a new life with very little in their hands.

Despite efforts and struggles, Sigaud's (1988) final analysis of family resettlement is overwhelming, using the term "military operation" to evacuate a territory. According to the data gathered by Silva and Germani (2009), based on the work of Duqué and Sigaud, a total of 11,853 families migrated, an estimated 60,000 people having to change residence and being forced to rebuild their lives somewhere else.

The Sobradinho Dam is emblematic of its location, in the context of military junta, in the heart of a territory occupied by poor rural populations. Although the Sobradinho Dam was not the first built on the São Francisco River, it reduced the authorities' scruples to install further large structures on the river, such as other hydroelectric plants in the following years, and more recently, the monumental works aimed at diverting the river flow to feed the rest of the Nordeste region.

Itaipu

The Itaipu Dam, built between 1975 and 1982, is the most famous Brazilian dam in the country and in the world. This can be explained not only by its size, as it remained for a very long time the largest dam in the world, but also by its international character because it is built on a border and therefore required the involvement of five countries before its construction could begin. The Itaipu Dam is built on the Paraná River, the natural border between Brazil and Paraguay.

The Itaipu Dam stands out for its unique character in international relations and international law. In his very comprehensive publications, Christian Caubet explains all the stages and negotiations necessary before the construction of the dam (1991). He stresses the complexity of the debates to resolve conflicts of interest between countries that wanted to exploit the same shared resource in different ways. The dam had to be built in an ancient zone of conflict, caused by the delimitation of the borders, notably on the site of "Sete Quedas" disputed by Brazil and Paraguay (Benetta 2002).

The Itaipu Dam, located on the Paraná River and in the Great Plata River basin (3 million km^2, half of which is in the Paraná basin), involved the five countries that have land in the basin. Brazil, Paraguay, and Argentina were the main actors in this project, while Uruguay and Bolivia, who also own land in the basin, played a minor role. While the first two quickly found an agreement for the construction of the dam, Uruguay, in a global context of military power in Latin America, entered the game of negotiations. The project came under discussion between Brazil and Paraguay in 1966 with the "Ata de Iguaçu" (Germani 2003), and a treaty of construction of Itaipu was signed in 1973 (Germani 2003). It was not until 1979, after long negotiations, that a tripartite agreement was finally signed (Caubet 1991).

For a long time, the dam was regarded as "the largest in the world," thus becoming a propaganda tool for Brazilian civil construction and military power.

In terms of installed capacity, the hydroelectric plant now takes second place with a potential of 14,000 MW. The construction work was huge in terms of workers (25,000), amount of concrete used, and financial resources (total cost of about 30 million dollars). It is now a recognized tourist destination in Brazil.

Due to its size, the social and environmental impacts of Itaipu are numerous. G. Germani, who studied this aspect, explained in her 2003 book how the project came about without taking into account at any time the interests of the local population and the life projects of the inhabitants of the region (2003). In total, an estimated 42,444 people were expelled from their land, 90% of them from rural areas (Germani 2003). The expellees were spread over eight municipalities, with over 110,000 ha flooded, which G. Germani described as an "area of conflict" (2003). She considers this lack of involvement as the starting point for the many struggles that appeared in the region, fueled in addition by the status of the inhabitants (owners or not) which profoundly changed the conditions of negotiations with Itaipu Binacional (Germani 2003), the company responsible for the construction of the dam.

G. Germani divides into three phases the struggles of the expropriated families which lead to the organization of the social movement.

The first phase, between 1973 and 1977, corresponds to the beginning of construction and the first stage of compensation. The population was informed of the construction of the structure but did not participate in any decision making. A first inventory of the flooded lands was made in 1973, and during meetings in 1976, the company gradually explained the notion of expropriation to families. There was still a climate of trust between the population, the company that promised to pay a fair price, and the government that supported the process more or less. That level of confidence however diminished because of the low compensation that was offered, the absence of official documents, the individualization of the process, the non-compensation of land for squatters (posseiros), pressures to accept compensation, etc. (Germani 2003). In this context and seeing the problems increase, the Catholic and Lutheran churches, who were direct witnesses in the communities, decided to get involved. However, a catalyst was still needed to unite the population who wanted to demonstrate but was afraid of the junta and its powers. In 1977, the Pastoral Commission of the Earth (CPT) officially established itself in the region and played a central role in the organization of the families expropriated by Itaipu (Germani 2003).

The second phase (1978–1979) corresponds to the time when families started to organize themselves. Two meetings were held to structure the group of demonstrators. The first, in October 1978, brought together 1500 farmers and concluded with a written summary of the problems. Signed by more than 1000 people, it was handed over to President Ernesto Geisel who was visiting the site the following week. The document had many consequences for the government, the Itaipu Binacional Company, and the press. Because of its disclosure, the church placed itself on the front line for the representation of the expropriated. At the beginning of 1979, the company increased the compensation value by 40% but this did not stop the public's dissatisfaction as the conflict kept spreading, notably because of the posseiros' situation. A second meeting, held in April 1979, was attended by more than 2000 farmers. This meeting marks the continuation of the protest against Itaipu but also against the National Institute for Colonization and Agrarian Reform (INCRA), which, according to farmers, did not play its role of helping the resettlement of expropriated farmers and was taking too long to legalize the newly occupied land. The pressure of bad publicity for Itaipu by the expropriated forced the company to find land solutions more or less appropriate (example of the "Agrarian Market," sort of internal market in the state of Paraná where families can find land to buy). Compensation was slow and distant and did not satisfy many of the expropriated. The pressure continued to rise, there was a risk of violence (Germani 2003), and most of the expropriated families were still awaiting compensation or a fair price.

The third phase began in 1980 with the establishment of a camp surrounding the headquarters of Itaipu Binacional in the municipality of Santa Helena. Due to its popularity, the camp grew in size with the arrival of farmers from all over the region. Through it was created the "Justice and Earth Movement," the first organized and peaceful social group aimed at bringing together victims of hydroelectric dams. The company quickly started to accept a lot of settler claims but not at inflation-adjusted prices. As a result, the camp and the demonstrations continued to strengthen and started to receive support from other regions of Brazil. After 11 days, the company accepted the adjustment but under certain conditions which the demonstrators did not find suitable. They decided to intensify their action by threatening to move the encampment toward Foz do'Iguaçu. This pressure proved effective and the negotiations resumed, finally leading to an agreement and the dismantling of the camp. The social movement nevertheless continued and organized an encampment of 56 days from March 1981. The pressure once again worked, and families gradually started to receive compensation, but the struggle continued for several months. During all these years of conflict, the main point of contention had always been access to land (Schmitt 2008), the first condition for the survival of farm families.

Lastly, another significant social impact affected indigenous populations in the flooded area. Guarani families living in the indigenous land (TI) of Ocoí (Santos and Nacke 2003) had two-thirds of their reserve flooded and were at the same time threatened by the resettlement of expropriated farmers. In the mid-1980s, 11 families, totaling 27 people, lived in this IL. In 1986, the indigenous people internationalized the conflict by sending a letter to the World Bank which succeeded in alerting the Brazilian authorities. As the National Indian Foundation (FUNAI) was in crisis (Santos and Nacke 2003), the social situation of the indigenous people did not improve but deteriorated with the arrival of other Guarani families, bringing the total population to more than 700 people (Terras Indigenas no Brasil 2002). The indigenous people had to move twice, on lands bordering the reservoir (Benetta 2002), but their living conditions remained precarious.

The dam has become an important tourist destination in the west of the Paraná state, and efforts have been made to remember the workers and expropriated people through various initiatives such as museum. Families who have been victims of the dam have managed to rebuild their lives after years of struggle.

Belo Monte

The Belo Monte Dam has been the most debated dam since the beginning of the twenty-first century recent years and has been observed with international attention. Built in the Amazon, the dam is one of the three largest in the world resulting in several significant impacts.

The origin of the construction project for the Belo Monte Dam dates back to the 1970s when the military authorities asked the Camargo Corrêa Company to carry out a study on the hydroelectric potential of the Xingu River. In 1980, while other large dams were under construction, the government was considering the construction of seven dams for a total production of 22,300 MW and a flooded area of 19,000 km^2 (Filho 2005). The project did not come to fruition, but the plans were retained by the Brazilian electricity sector's decision makers.

In 1987, the project was revised to create the Altamira Complex with two dams (Kararaô and Babaquara) with an installed potential of 17,700 MW. The project quickly spurred several demonstrations. The national context of fight against dams (see works by Filho (2005)), the economic context (refusal of financing by the World Bank in 1990), and the political scenario did not favor construction. Once again, the project was placed on hold.

The transition to the twenty-first century proved to be decisive as the project returned under the name of Belo Monte and joined the Growth Acceleration Program (PAC) in 2004. The project was downsized compared to previous revisions, with a flooded area down from 1200 to 400 km^2, thus reducing its impact on indigenous reserves in the Altamira region. The Belo Monte Dam has a potential power output of 11,000 MW. This rated power output has often been questioned because it is only really achievable during the rainy season which typically runs from January to June.

The construction of the Belo Monte Dam finally started and from then on could not be stopped. It quickly appeared as very high on the agenda of the political powers in place. In an interview in April 2010, the Brazilian President Luiz Inácio Lula da Silva said that if private companies were not interested in building Belo Monte the country would build it alone. The justification for the project was twofold: fear of an "apagão" (general power outage in the country), which had been used as a justification by the sector since the last power outage in 2001, and the discourse on the need to build dams so as not to slow down the development of the country. The projected cost of the project (30 billion reais or 9 billion of dollars) was a real deterrent for private companies, but the government played all the strings to encourage the private sector to invest. The state also took part in the consortium (Norte Energia) chosen for the construction and operation of the plant. A group made of Eletrobras, CHESF, and Eletronorte currently owns 49% of the plant, while the other minority shareholders are Cemig, Vale, Light, Neoenergia, Petros, Funcef, J. Malucelli Energia and Sinobras. It should be noted that Norte Energia has been on sale since early 2017 and that its future remains uncertain in particular because of investigations for corruption in the framework of the national survey "Lava Jato."

In terms of funding, the government assumed the risks associated with the project. Eighty percentage of the project was funded (Brzezinski 2014) by the National Bank for Economic and Social Development (BNDES) which offered special conditions (interest rates and payment terms) to the government. In return, the bank received payment insurance because the government guaranteed itself by the public treasury and pension funds. Once the financial and economic package had been put in place, the government was committed to defend the project at other levels.

The Belo Monte project could not be hindered by environmental authorities. It did not seem to follow the normal process of environmental licensing after the withdrawal of a president of IBAMA (Sakamoto 2011). A group of nearly forty academic researchers gathered in 2009 to denounce the problems and shortcomings of Belo Monte's environmental impact report (EIA) in a 230-page document (Magalhães and Hernandez 2009). Even though the document was widely disseminated and relayed to the media, it had limited consequences for the finalization of the Belo Monte project.

The government also faced opposition from the media, NGOs, and celebrities, well-known nationally (such as known actors from Globo, the largest television network in Latin America) and internationally (actors like Arnold Schwarzenegger and singers such as Sting). This buzz against Belo Monte was only temporary, and public opposition mostly disappeared once construction started. NGOs nevertheless continued their fight against the project but, with work progressing, opted to focus more on the respect of environmental rules and the social conditions of the families affected by the dam.

The government also had to respond and defend itself before the courts. In 2005, the Minister Chief of the Civil House defended the project before the Supreme Federal Court. The judicial conflict then focused on the issue of environmental licenses (Fleury 2013) involving the Federal Supreme Court, the Public Prosecution Office, regional courts, and civil society organizations.

Finally, the government also defended the plant construction against the local populations by using deterrence. In 2013, the National Security Force was sent to Altamira for 90 days in order to suppress protests and guarantee the continuity of the construction (Brzezinski 2014). The government therefore used all the state apparatuses at its disposal to guarantee the construction of the Belo Monte Dam.

The construction of the Belo Monte Dam was at the center of a major environmental conflict involving the public authorities, private authorities, the media, and civil society and its representatives. All layers of Brazilian society were involved in this conflict, and its repercussions were so wide that it was almost necessary to review the concept of environmental conflict (Fleury 2013). The conflict was so large and complex that, in an analysis of the project's vision on social networks, there was a clear division of opinions for and against the project (Santos and Mello 2014), thus demonstrating the lack of discussions and democracy in the construction of the project.

The Belo Monte plant is characterized by the scale and diversity of its impacts. The entire region has been affected by the construction of the dam whether it is the neighboring municipalities, the Trans-Amazonian road, the town of Altamira and the Xingu meander or Volta Grande. Norte Energia defined a direct area of influence which consists of five municipalities near the dam (Altamira, Anapu, Brasil Novo, Senador José Porfírio, and Vitória do Xingu) and an indirect influence area consisting of six municipalities (Gurupá, Medicilândia, Pacajá, Placas, Porto de Moz, and Uruará).

All the municipalities in the region have been affected to a certain extent, at least in their demographics, mainly because of the attractive opportunities offered by the construction of the plant. Local municipalities received various subsidies and compensations from the company.

The Trans-Amazonian Highway, which passes by the foot of the main powerhouse of Belo Monte Dam, crosses an agricultural area occupied by settlers since the 1970s and 1980s. The influence of the dam, whether positive or negative, has been felt. There has been a renewal of the ownership of agricultural lots with the arrival of populations in the region and also a new shortage of labor for agricultural activities. In addition, the road took advantage of the construction of the plant to increase its asphalted surface thus facilitating access and transport in the region.

The city of Altamira is the largest urban area having been affected by the construction of a hydroelectric plant in Brazil. With just over 75,000 inhabitants in 2010 (IBGE 2010a, b, c), Altamira is the logistical and nerve center of the Belo Monte region but also a city along the Xingu River which saw its water level rise with the construction of the dam. According to RIMA (Eletrobras, Eletronorte and MME 2009), 4362 families living in Altamira moved in 2009 for a total of 16,420 people, and according to a 2016 communication from the company, 30,000 people benefited from new houses in Altamira, about twice what was expected. Other structural changes accompanied migration in Altamira: a lack of health infrastructure, a rise in the price of land, and above all an unprecedented level of insecurity. In 2015, Altamira became the most violent city in Brazil with a homicide rate of more than 105 deaths per 100,000 inhabitants (IPEA 2017), while the national average is 28.9.

The final large area affected by the Belo Monte Dam is the Volta Grande meander, downstream of the first dam of Belo Monte. This part of the Xingu River is now isolated from the rest of the river because it has been cut by the dam and the floodway of Belo Monte. The flow of water has therefore been greatly reduced with negative consequences on the fauna, flora, and people living along this part of the river, about a 100 kms long. The Volta Grande hosts numerous "*ribeirinhos*" on the shores and islands, a community of artisanal gold diggers (about 300 people) and two Indian reserves: Paquiçamba and Arara da Volta Grande do Xingu. According to a census of the Ministry of Health of 2014, they count 238 indigenous people (respectively, 95 and 143), on 30,000 ha. The flow of water running into the Volta Grande after the construction of the dam has greatly diminished and does not allow indigenous families to meet their needs for food and transport, etc. (Bermann 2012). In addition, a major project to open a gold mine is also in the pipeline:

the Belo Sun project. Announced as the largest open-pit mine in Brazil, it would involve building a tailings dam (waste) on the edge of Xingu.

The riverain families are the most affected by the dam, and social and geographical circumstances make it difficult to measure its full impact. First, it is very hard to locate them; riverains pretty often have two homes, one by the river and other in town, and this lifestyle is not accepted by the authorities. Reparations to the riverain families have generally been paid in money; therefore, they did not take part in the relocation effort and went off the radar of official programs that track the migrants. So, when it comes to mobility and reparations, it is as if they have disappeared from the analysis after the building of the dam. It is important to notice that these families suffered a lot with the change of residence, especially when it comes to their livelihoods, which have always been connected to the river. In most cases, these families were unable to overcome the loss of their natural environment, as can be seen in the analysis made by the SBPC in 2017. The five riverain groups identified by the research were deprived of their means of production and left with no possibility of rebuilding their riverain lifestyle. Because their new houses were not built on riversides, their traditional life space, so essential to their existence (Cernea 2000), was destroyed, and this led many riverains from Belo Monte into a very precarious situation. These are the groups most negatively affected by the dam, left in a "critical situation of violence, social, environmental, and alimentary insecurity" (Magalhães 2017).

Other impacts have affected the Altamira region but were less substantial and more isolated, for example the changes encountered in terms of fishing in the mining reserves upstream of the Xingu River. Traditional populations especially indigenous people that had been affected in different ways by the construction of the dam took part in the fight against the construction as described in the book of Fillho of 2005 "Tenotã-Mõ." Native American populations were deeply affected by the building of the Belo Monte Dam, especially those living downstream: the Juruna, from Paquiçamba, the Arara, from Volta Grande do Xingu, and the Xikrin, from Bacajá. In Volta Grande do Xingu, besides the impact on the land, the greatest impact was the dismemberment of groups and communities. Amid pressures, negotiations, political, and territorial games, the Native American groups did not enter any agreement and got divided. Thus, the number of villages went from 19 to 39 between 2010 and 2015 (Fearnside 2017a). Once again, Native American groups pay the price of development (Fearnside 2017b): Even with all the struggles before the building of the dam, they have been affected by deep local changes.

Urban centers near Belo Monte have also been affected by the construction of the dam such as those of Brasil Novo and Vitoria do Xingu which saw their population increase. Economic activity in the region was booming at the beginning of the construction of the dam but quickly came down.

These four examples of hydroelectric plants in Brazil give an overview of the situations that accompany these constructions and the diversity of their consequences. The role of the public authorities in the implementation of projects and the almost automatic negative consequences for the populations can be noted. We will describe in the next chapters the impacts and adaptations of the victims of dams in Brazil.

1.2 Scales Analyses

1.2.1 Local Scale

The issue of proximity must be taken into account to understand the territorial and spatial effects of dams in Brazil. Depending on the perspective taken, transformations can indeed be seen differently, therefore influencing the comprehension of the issues. The effects of dam constructions are more or less foreseen by preliminary environmental studies as they have a ripple effect, with initial transformations producing secondary modifications. We will also talk about the impacts which are seen as consequences of the effects. Effects and impacts are inherent to dams, but they are more or less controlled by companies. All dams affect their area of influence on various scales, with intensities varying with distance. In addition, the larger the dam, the larger the impact analysis required. The analyses mainly target demographic, economic, cultural, environmental, social, and infrastructure issues. For each of these categories, it is possible to describe the type (positive or negative), mode (direct or indirect), intensity, duration, permanence (reversible or not), and zone of direct and close influence of the impacts which can then be used as the first analysis criterion.

The impact study must first focus on the local level, which is the area of influence closest to the dam, plant, and reservoir. The local area of influence covers a radius of a few dozen kms around the construction site.

The local effects are determined by looking at the different parts of the dam. The first area affected by the construction is the site chosen to build the plant and the dam. The site of the reservoir only gets affected after completion of the first stage of construction.

The choice of the construction site is of the utmost importance due to the changes induced by the reservoir and dam as a whole. There are several examples of projects' relocation following protests and re-evaluations. For instance, the projected sites for the Belo Monte and Machadinho dams had to be moved to greatly reduce their future effects. Some engineers who think and plan the dams only have a selective view of the final product, a technical view mainly focused on the production of energy. They sometimes ignore other aspects of the project such as environmental concerns, impacts on local populations, historical value of the site. The dam location is therefore often discussed in the early stages of planning because it directly influences the area and intensity of local impacts, whether positive or negative.

All dams have local impacts, whether big dams like Itaipu or smaller ones of which there are many in the south and southeast of the country. The succession of smaller dams along the same river also has cumulative effects on local impacts. Global change then takes on another dimension.

The first systematic effect of a dam construction concerns the indigenous population. As Brazil's rural areas are not deserted, the construction of a dam systematically results in the displacement of rural families living nearby

(84% of Brazil's population lives in urban areas according to the IBGE criteria, during the census of 2010). The local impacts of dams are mainly experienced by families living on the shores of the river where the reservoir is created and in the dam and plant area. Dams are almost always built in rural areas, and it is difficult to imagine a dam near a city. Brazil's rural areas, though not totally neglected, are areas of relative demographic decline, and riverbanks are abandoned by agribusiness capital and thus occupied by small farmers and/or fishermen.

A small dam like Monjolinho in the Rio Grande do Sul moved between 200 and 400 families, while a large dam like Sobradinho mobilized about 60,000 people. The Monjolinho Dam, despite its small size (74 MW installed on the Passo Fundo River), has significant local impacts on several levels. From a demographic point of view (information available on the portal of the Socio-Environmental Observatory of Dams), according to the Movement of People Affected by Dams (MAB), about 400 families had to leave their home, whereas according to the preliminary environmental study only 217 families were to move. Indigenous communities were also affected by the dam (Kaingang and Guarani) and received special compensations. Families living on the banks of dammed rivers always experience the local impacts of the dam.

The local populations most rapidly affected by a dam are often fishermen because their main resource, the river, gets disrupted from the early stages of construction. Alterations made to the river affect their economic activity and challenge their lifestyle and livelihood. Local economic consequences may not be significant in terms of value, but they profoundly impact local families and economies which are often already fragile. A large influx of capital in a short period of time changes local markets as demand increases and diversifies. Local people have nevertheless tried to respond. For example, farm families sometimes abandon their initial activity to get into something else hoping to find a larger source of income but forgetting the risks associated with this change and the limited period of this exceptional demand. The local economy is thus positively impacted for a relatively short time, and then it takes some time for the situation to stabilize once the initial excitement of dam activities has passed.

Dam constructions also have other automatic consequences, such as at the environmental level. Hydroelectric plants are often presented in Brazil as renewable and "green" energy sources, but the environmental impacts of their construction are irreversible. Damming a watercourse and building a power plant inevitably change nature. Even though the transformations do not necessarily take long, as they only last during the time of construction, their effects are permanent. A new environmental order then takes its course. The same applies to impacts on the landscape, whether real or imagined. The landscape as it was before the dam is destroyed and disappears little by little to give way to a new and more complex landscape integrating new elements.

An important local benefit concerns the collective infrastructures which are modified with the development of the site. It is indeed common to see improvements in transport infrastructures, mainly roads to be able to transport machines and heavy materials to the construction site. Investments are also sometimes made at airports and bus stations to facilitate access for the people coming to work on the site. Health and safety infrastructure (police) improvements are also often found as they directly serve the interests of the companies constructing the dams. These infrastructures also benefit the local population. Nevertheless, they only arise for the benefit of the dam and no investment is made beyond that. The dam of Foz do Chapecó provides a very interesting example: A road was asphalted in the municipality of Águas de Chapecó (state of Santa Catarina) as this was the main access to the site, but, on the other side of the river, the municipality of Alpestre (Rio Grande do Sul) did not see any improvement and kept a dirt road between the center of the municipality and the dam. The road improvement was therefore only carried out to assist in the construction of the dam, creating a striking contrast with the rest of the area.

The level of local impacts is different from the other impact zones due to the time factor. Everything that happens at the local level seems more intense and faster. The arrival of the dam causes modifications that mark the local space more deeply for a specified period. From a local demographic point of view for example, there is an influx of people even before the start of construction, with families settling to find work. During construction, agitation reaches its peak, and once the construction is completed, almost everyone leaves and only a part of the population present before the arrival of the dam and some people who decide to settle there remain.

The local effects of the establishment of dams in Brazil can sometimes be considered brutal and frenzied. Their repercussions are diverse (demographics, economy, environment, etc.) and deeply mark the local territory of the dam because they disrupt a rural "peace." The local impacts of dams are undeniably greater than for at any other level.

1.2.2 Regional Scale

To understand the effects of dams at the regional level, it is first important to define the region as the area around the dam which gets affected either by the construction process or by the consequences of the installation. A functional definition of the region conveys the idea of internal cohesion and homogeneity of space. Administrative or border regions are not taken into account here because dams are built on rivers that often serve as boundaries, thus bringing together various administrative areas or sectors into one unit.

The notion of region refers to areas more or less directly affected by dams. Not all dams and their hydroelectric plants have a regional influence: The smaller ones indeed only affect the various sectors of society locally, and the region around them does not feel the consequences of the constructions. But a group of small dams, built in a row along a stream, might have greater influence and, therefore, reach an entire region.

The most significant regional impacts are related to the economy. The construction of a dam can revitalize the whole area around the site. Economically, the region is measured by the intensity of exchanges between a center and various peripheries. With the creation of a dam, exchanges of all kinds multiply and accelerate for a specific period of time. The region becomes an area of asymmetric interactions where flows of goods and services intersect. Regional relations may be more or less important and intense depending on the project.

A study of the workforce allows a better understanding of the influence a dam may have on the regional economy. The construction stage requires a large amount of manpower, mostly non- or low-qualified labor. At the very beginning of construction, the area near the dam is full of people looking for employment opportunities in the hope to improve their living conditions and incomes. The creation of a dam almost systematically triggers migratory flows of population in search of work in the south, Nordeste region, and north of the country. The workforce often depopulates rural areas and nearby urban areas to temporarily settle closer to the site.

The regional economy also benefits from the construction of a dam through the services. The urban centers in the vicinity of the site become the logistical hubs of construction. Services and businesses such as hotels, restaurants, transport, communication, work to meet the growing demand of the population. Urban centers in the nearby region quickly get occupied by migrants coming to work, and service prices change rapidly. For instance, the town of Altamira, 50 kms from the Belo Monte Dam, saw a very sharp increase in rent. The regional economy experiences a real explosion with all entrepreneurs trying to take advantage of the situation in one way or another.

This is also true of the informal economy and illegal activities that accompany dam constructions in Brazil, with a wide variety of sectors not legally recorded. With the massive arrival of workers on construction sites, some activities spontaneously arise to meet the workers' needs and suit their incomes, sometimes quite low and irregular. These informal businesses commonly include bars and nightclubs built in a hurry, documentation services such as photocopies and photograph services, taxi and motorcycle taxi services not legally registered with the authorities. These could be legally authorized but for lack of time, structure and various other reasons (money, paperwork, tax, etc.), entrepreneurs often decide not to go through this step in order to earn money as quickly as possible.

All sorts of imaginable illegal activities can be found around dams: drugs, prostitution, sale of arms, etc. Cases of prostitution and child prostitution have been revealed around various dam sites. The Pastoral Commission of the Earth (CPT) notably denounced crimes of prostitution and pedophilia around the Jirau and Santo Antonio dams on the Madeira River (Comissão Pastoral da Terra de Rondônia 2011). The organization explains these issues by the large flow of population in the region because of the projects and the presence of the border with Bolivia only a few dozen kms away. Another case in the Belo Monte Dam area was also exposed in the media where police discovered a club involved in prostitution, including teenage prostitution, and bonded labor (Glass 2013; Leite 2015). The climate of violence surrounding this case is indicative of the human relationships that may exist around the construction sites.

There are also cases of employment under slavery-like conditions around certain hydroelectric plants. The CPT, known for its substantive work on the denunciation of bonded labor in Brazil, reported 38 cases during the construction of the Jirau Dam on the Madeira River in the Amazon in September 2009 (Rossato 2013). The Regional Labor and Employment Superintendency of Rondônia and the Ministry of Public Labor Prosecution then released 38 people working in slavery-like conditions for the construction of new quarters aimed at accommodating future-affected families. Without wanting to extrapolate too much, it is possible to imagine that this type of situation still exists for other dam constructions.

Regional infrastructures linked to the smooth running of the economy also come under pressure during the construction of a dam, such as airports, roads, railway stations, hospitals, health facilities, police stations. Their overuse sometimes causes system breakdowns. For example, as of 2011, the health services in Altamira became overloaded and could no longer cope with the influx of patients. Regional infrastructures are generally marked by the effects of a dam construction much more intensely than at the local level. This is a sign of an increased concentration of services around a few centers in rural areas and in nearby urban centers, increasing pressures and constraints on equipment. The same phenomenon took place near Altamira, in Tucuruí, during the construction of the dam at the end of the 1970s. The construction attracted so many people that the population increased sixfold between 1970 and 1980 (IBGE Census 1970 and 1980), rising from just under 9,000 to 61,100 in 1980. The construction of the Sobradinho Dam in Bahia also saw the population almost double between 1970 and 1980 when the dam was built. It increased by 57,000 people between the two censuses (municipality of Juazeiro). These demographic "booms" weigh heavily on the infrastructures and disrupt the regional dynamics.

The regional area is also marked by the consequences of dams with the forced migrations of families living near the river. *Atingidos* (affected) are obliged to find new residences, but despite their wish to remain close to their former home, they have to move to more remote areas (urban areas) within the region.

This flow of families and the search for agricultural land exerts a new pressure on land ownership. Increased demand automatically raises the prices of available land but also encourages owners to sell their property. Intermediaries benefit from this new momentum.

The significance of changes at the regional level is due to a cumulative process. The accumulation of development works in a region indeed multiplies impacts, both positive and negative. This is illustrated by the link between dam and aluminum smelter or mines. The phenomenon also exists when several dams are built in the same catchment area as in the case of the Uruguay River or the Paranapanema River which counts 11 hydroelectric plants along its 920 kms.

Various effects are more intense at the regional level than at the local level, and these durably mark the territory. Some dams take on a fundamental role at the regional level, even becoming the symbols of those spaces. They can become a hub for the region, therefore creating a real focal point for the area. The region is often conceived as an integrated subspace with a strong homogeneity. It is a space of cohesion where flows are organized around a common characteristic which can be the influence of a hydroelectric dam.

1.2.3 Global Scale

The analysis of the national and international effects of dams requires a broader view and a step back to look at the conditions of their construction in Brazil. All Brazilian hydroelectric plants are part of a national framework and in some cases international frameworks too: They respond to a national energy policy and to the planning of the electricity sector. But not all dams have the same significance: The smaller ones only have little influence, and their repercussions remain at the local and/or regional level.

There are not many dams with both national and international implications in Brazil, but the few that there are often very well known. This is notably the case of Itaipu, Belo Monte, Foz do Chapecó, Itá, and Tucuruí, among others. These dams are all trademarks and symbols of the Brazilian hydroelectric production, each for a particular reason. Their repercussions however can be of varying intensity and scale: Belo Monte has been the most discussed dam since 2010, while Itá retains the reputation of a well-known dam whose reservoir flooded a whole city, leaving only the very top of the church visible.

The largest dams are often the result of political plans and projects that go beyond simple electricity production. Thus, the Sobradinho, Itaipu, and Belo Monte dams were built because the Brasilia government had strongly supported, in one way or another, their construction. The Sobradinho Dam, built in the period of the military dictatorship, is described as very rigid in its planning and its construction, and authoritarian in its relations with the local population (Sigaud 1988). Dams serve as an example and symbol of the political actions of the governments and demonstrate the power of the central state. For instance, the Itaipu Dam illustrates the strength of the military government which overruled local and regional political powers (state of Paraná) on various occasions (Germani 2003). During the military dictatorship, large dams served as a showcase for the power in place, giving an image of power, order, progress, development, etc. The military period was over 30 years ago, but similar political power plays still exist.

Today, democracy has returned but major development works continue to be used as symbols and marks for politicians. The construction of the Belo Monte Dam was, for example, based on the will of some of the most influential leaders of the Workers' Party. It shows the power of central government, in a similar fashion as during the military era.

Decisions to construct dams are made by the Ministry of Mines and Energy (MME) following consultations and impact reports. But the final decision is taken before the reports and the public consultations, which are then used to legitimize the whole process. However, the final decision is often unilateral and centralized

(Boanada et al. 2016), leaving little room for the opinions of local and regional populations. The decision comes from the top, at the national level, for a grassroots application at the local level.

Some dams have even more influence when they are built on international borders as is the case for Itaipu. As previously explained, international negotiations between five countries were required before its construction could begin. Built between Brazil and Paraguay, complex negotiations took place notably with Argentina, as described in detail by Caubet (1991). C. Caubet stressed that diplomacy and negotiations were always on the agenda despite tensions, and even asserted that the regional controversy of the construction of the dam had repercussions in international institutions, notably for legal matters, including International Environmental Law (1991). Negotiations are needed to deal with national economic effects, but also with the management of resources. The catchment area of rivers that serve as a border is often very large, and the governments concerned sometimes have different priorities.

National and international implications must be addressed with the question of funding of dams. Between 1950 and 1980, funding was often provided through state and international agencies, such as the World Bank, the International Bank for Reconstruction and Development (IBRD), or the International Monetary Fund (IMF). For instance, the construction of the Furnas and Sobradinho Dams in the 1970s was financed by the World Bank. These funds have however been greatly reduced (see Equator Principles) and are now lacking for some players in the Brazilian electricity sector (Goldemberg, in a 2008 article in the Estado de São Paulo newspaper).

With the privatization and restructuring of the sector in the 1990s, the financing of large dams changed rapidly. The government withdrew from direct financing, and investments started to come from the private sector. Currently, a large part of financing for development works such as dams comes from the National Development Bank (BNDES), an autonomous public bank (Uderman and Calvacante 2011). Already present in the middle of the twentieth century, its role has been transformed to support new construction projects in the context of privatization. From a global point of view, funding in the hydroelectric field has decreased in comparison with the 1970s and 1980s but remains important for this sector which is fundamental for the development of the country. Between 2000 and 2004, the BNDES provided almost 20 billion reais (6 billion of dollars) to hydroelectric generation projects (BNDES 2004) like those of the extension of the Tucuruí Dam or the construction of the Itá Dam. Major current projects in Brazil get almost all of their funding from the BNDES, such as the Estreito, Santo Antonio, and Belo Monte dams. The BNDES plays a fundamental role in financing investments in the Brazilian electricity sector (Uderman and Calvacante 2011), but these have also been reduced as the bank wants to support other renewable energies (2016). The resources (water) become merchandise in lot of cases (Zhouri and Oliveira 2008).

Some large dams have become symbols not only for the political power, but also more generally for the population and society. All Brazilians know the Itaipu and the Belo Monte dams. These structures have become part of the country's history, national identity, and image. While most Brazilians are not really aware of the impacts of large dams, they know of their existence and they take pride in the technical and architectural achievements of the projects. Large dams serve to give a positive image of the nation. The Itaipu Dam is known around the world, its Wikipedia page is translated into more than 40 languages, and it is a popular site for tourists who come to visit the nearby Iguazu Falls.

At the same time, some dams have attracted the negative attention of international environmental protection groups (mainly NGOs), such as Greenpeace and the WWF. The latter two have officially opposed dam projects in Brazil and have a unique media power to attract global attention to these issues, especially when projects are to set up in the Amazon. Often associated with the protection of Amazonian indigenous peoples, these NGOs broadcast their messages internationally and can have a significant influence in Brazil.

With regard to hydroelectricity production, Brazil is a world leader, being the third largest producer, according to the International Hydropower Association (IHA 2017). Brazil's hydrogenerating capacity is just under 100,000 MW, well behind China's hegemony (more than 330,000 MW) and just behind the USA 102,000 MW. In 2016, Brazil was second in the world in terms of installed capacity increase, with an increase of 6300 MW, again behind China which is investing heavily in this sector (plus 11,700 MW in capacity in 2016). We can see on the following map the weight of each country in the global production of hydroelectricity, the impressive size of Brazil in the world (2006).

China is also an increasingly recurrent investor in the Brazilian electricity sector and thus in hydroelectric dams since 2013. Through the China Three Gorges (CTG) Company, one of the world's largest energy companies, China is positioning itself in the sector. CTG first acquired the rights to operate two dams on the Parana River (Jupiá and Ilha Solteira) before buying the Brazilian subsidiary of Duke Energy. As a result, CTG has achieved an installed production capacity in Brazil of 8270 MW (CTG 2017) and, in its own terms, has become "the second largest power generator in the country with private capital." CTG has a stake in 15 hydroelectric plants in Brazil and is in the midst of expansion, having just, in August 2017, signed a technical exchange agreement with Itaipu. Finally, it must be noted that Chinese investments in the sector are multiple with, at the same period, an agreement signed with the Chinese State Grid Company for the construction of a transmission line between Belo Monte and the state of Rio de Janeiro (2000 kms). With Chinese investments in the Brazilian hydroelectricity sector only just starting, one will have to wait to be able to fully analyze what seems like a shift in the history of electricity in Brazil.

To conclude, the effects of dams at the national and international levels involve larger and more comprehensive issues than at the local and regional levels. National and international effects are often less visible or less tangible. At the national level, they touch on imagination and politics. Some large dams have a significant impact on the history and collective memory of the country. Itaipu and Belo Monte, for example, are two important markers of the country's history. They also represent a political period, notably the military junta and the start of twenty-first century.

The construction of a dam is an event that can leave historical traces. For instance, the anecdote is that on the day of her dismissal, President Dilma Rousseff inaugurated the Belo Monte Dam, a controversial work she personally supported and which marked her government.

Internationally, dams create many global issues that affect the economy, finance, law, and so on. The Brazilian national economy is involved in the construction of dams, with financing, but also in the final production and electricity distribution to companies and individuals, and finally in the area of international investments, which have become increasingly open to foreign countries from the end of the twentieth century. The international effects of dams in Brazil have taken a new turn, the last 20 years, notably because of the recent interest of Chinese investors in the sector.

1.3 Landscapes and Territories Modifications

1.3.1 Landscape Evolution

The construction and dissemination of hydroelectric plants throughout Brazil end up changing the landscape. All regions and states of Brazil host dams and other large-scale structures with localized effects.

Widely studied by various geographical schools since the origin of the discipline, the notion of landscape has evolved a lot and is no longer limited to the rural landscapes traditionally analyzed. Similarly, the dichotomy between cultural and natural landscapes is often blurred in a kind of broad landscape shaping. We will use the notion of landscape as defined by Berque (1995): "Invested by human values, the landscape is at the same time a living environment, a heritage, a resource and a badge of identity." The landscape is the result of the interaction and combination of biological (physical and chemical) and anthropogenic elements, creating an original but constantly evolving scene. To read and understand a landscape, we must remember the importance of the human outlook, the potential interpretations, and the imaginary images. To identify how large dams change Brazilian landscapes, it is therefore essential to consider this geographical concept in all its complexity.

We will start by looking at the type of landscape and geographical structure in which dams are constructed. Rural areas are always a favorable environment for construction, as they are old and complex, and generally occupied by agricultural activities and human habitats. In Brazil, dams are present in all regions so they are established in almost all types of possible geomorphological, climatic, and biological conditions. They are found in the Amazon rainforest, in the savannah of the Nordeste, in the Southern Pampa, in the hills of the Atlantic Forest of the Sudeste coast, in the semiarid areas of the Nordeste, etc. Dams are built in landscapes that have already undergone anthropogenic interventions, even in the most isolated areas.

To better understand the effects of the construction of a dam in a landscape, we will use as an example a photograph of the construction site of the Belo Monte Dam, in the Northern region of Pará, in the Amazon. This photograph gives us a lot of information on the landscape and its transformation. It can be divided into two parts: The first is composed of the dam construction site and surrounding work areas, while the second encompasses "natural" elements such as the Xingu River, symbol of the Amazon rainforest, and the vegetation marked by the presence of cultivated fields, evidence of human presence. The photograph can nevertheless be interpreted in other ways, by taking into account the evolution of the landscapes.

The construction of a hydroelectric plant brings with it a number of new landscape features varying in size, intensity, speed of development, etc. These changes all mark the territory and the landscape. As a dam-sized structure involves the construction of a concrete wall several dozens of meters in size in a rural landscape, it profoundly alters the local landscape. The dam, the plant, and the reservoir cause significant, long-lasting, and profound transformations. The building process can be likened to a violent act, for, in the space of 3–5 years, a complete landscape gets permanently altered by the construction of an imposing civilian work.

Figure 1.2 illustrates how the landscape profile changes with the arrival of a dam. The dam delimits two distinct spaces, two landscapes: The dam and the plant are very localized spaces, whereas the reservoir, above the major bed of the river,

Fig. 1.2 Photograph of the construction site of the Belo Monte hydroelectric plant. Vitória do Xingu/PA/Brazil. Guillaume Leturcq, May 2015

covers a wider area. To analyze the effects of dams on the landscape, it is therefore necessary to differentiate the various areas between the dam and its reservoir as the observations will differ depending on the affected area. Figure 1.2 only shows the lower part of the dam, the site where the turbines are installed, but not the reservoir.

The impacts of dams on landscapes are twofold, both a transformation and a disappearance. Transformations concern the natural environment. The natural biological life continues but is altered by the dam and its plant, therefore modifying the environment. The lake presents a different natural dynamic than the river, and therefore, transformations in the environment only tend to appear at the end of construction. A project of this scale also puts pressure on the characteristics of the site and can alter its identity. Disappearances concern parts of the material heritage (physical and biological for example) and intangible heritage (culture, memories, etc.). A dam and its reservoir can destroy physical elements such as waterfalls, villages, forests, and they can also destroy the soul of a place, its esthetics, its culture, its memories, and so on. Changes in landscapes can therefore be visible and invisible. The historical landscape which existed before the dam was constructed gives way to a new complex post-dam landscape.

In the analysis of the landscape modifications, one must also take into account perceptions and images, and not only visible aspects. Beyond modifications and disappearances, various intangible elements accompany the construction of a dam. A dam conjures up images of force, power, technology, domination, development, modernity, etc. Each observer or inhabitant has their own interpretation, which is not necessarily based on the images often associated with dams which are disclosed and used by the people involved in the dam development. Other people may have a more critical view of the works and associate them with images of authoritarianism, destruction, expulsion, etc. Thus, the vision and images associated with dams obviously depend on the point of view of the observer and the elements of interpretation at their disposal.

The people who observe the landscapes before and after the construction of a dam do so with their values, their history, their ideological position, and their role in the situation. The readings are therefore fundamentally different between a person forced to change residence and an engineer responsible for installing turbines. During an official visit of Belo Monte, we observed the opposition of landscape readings for the site. Our guide, who had been an engineer in the electrical sector for over 30 years, offered a look of admiration for the scale of the worksite, the means deployed, the size of the dam, and the impression of grandeur it gave. Among the visitors, there was also a student in ecology and sociology who was discovering this type of work for the first time and was outraged by its presence in the Amazon. She was astounded by the extent of the construction site and the scale of the potential upheavals on the environment. The effects of dams alter the vision people may have of it and therefore modify the image they have of its imprint on the landscape.

The perception of the landscape transformations around the dams varies with time and the duration of observation. The initial shock caused by the construction of the dam greatly affects the reading of the landscape, but, over time, the negative characteristics tend to fade and observations relative to the history of the structure and its integration into the local environment "soften." For example, dam reservoirs are often better accepted visually and for their utility in the landscape after a few years, once new uses have been found for them. It seems that the negative view of the dam tends to diminish over the years. For example, the landscape around the reservoir of the Itá Dam, with the top of the old church of the Sunken Village, greatly revolted the victims of the dam during the rise of the waters but later became the symbol of the dam and of the city.

The irreversible impacts of dams on the landscape effectively cancel out the arguments that present hydroelectric plants as "clean" sources of energy. Often presented as such in Brazil, as a means to justify how many there are, dams mark forever the landscapes and transform the geographical spaces, more so than other human interventions. The reservoirs of the Sobradinho and Balbina dams, respectively, covering, officially, areas of 4200 and 2400 km^2 in the semiarid Nordeste and in the Amazon, are the most striking examples in Brazil (Cf. Figs. 1.3 and 1.4). The irreversible and brutal impact of building a dam permanently affects the landscape.

After observing their extensive impacts, some countries are beginning to dismantle dams after their recommended lifespan has passed or because of a change of direction in energy policy. This has notably happened in North America

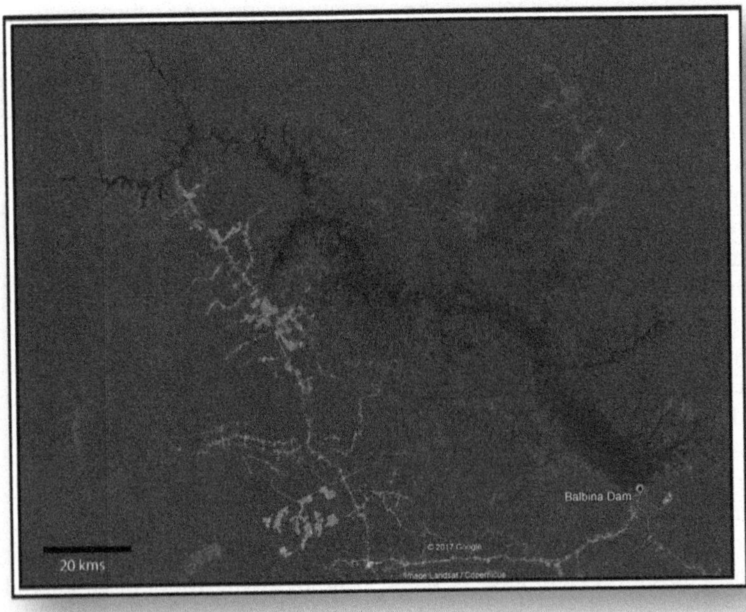

Fig. 1.3 Image of Balbina Dam reservatory. From Google Image, 2017

Fig. 1.4 Image of Sobradinho Dam reservoir. From Google Image, 2017

(dam destroyed in Washington State in the USA on the Elwha River for example) and in Europe (Poutès Dam in the Haute Loire region in France), and we may wonder whether this could potentially also happen in Brazil in the future. This however still seems fairly unlikely for a country that still relies a lot on hydroelectricity. It should be noted that these dams being dismantled were of minor importance in terms of electricity generation and one will have to wait a long time before seeing a complete reversal of the situation.

Finally, some dams try to blend into the landscape or at least to limit their visual impact by using a variety of techniques and illusions, such as planting vegetation on the wall of the dam or painting it in similar colors as the local landscape. Such a method was used for a dam linked to a deactivated mine (Akinaga et al. 2010), but this has not yet been done for a hydroelectric plant in Brazil. In Uruguay, part of the dam of Rincón del Bonete has been vegetated in order to try and make it blend into the landscape.

To conclude, hydroelectric plants significantly mark the landscapes in which they are located, modifying the very definition of the places, and bringing a new appearance and a new identity to the area.

1.3.2 *Territories' Evolution*

The building of hydroelectric dams in Brazil is changing the territories. The term "territory" is a popular word which is however sometimes misused, so we will use the definition of Guy Di Meo, French geographer, for whom the territory "testifies to an economic, ideological, and political appropriation of space by groups that give themselves a particular representation of themselves, of their history, of their singularity" (Di Méo 1998). He also emphasizes that geographical analysis, even more so in social geography, is concerned with "relational and multidimensional territorialities." In the case of dams, it is necessary to analyze the changes and evolutions of the territories by first considering the double phenomenon of insertion/exclusion of the populations on the territory and then by presenting the roles of the territorial development stakeholders.

To understand the inclusion and exclusion of populations in spaces and territories, it is essential to look at the notion of migration. Most dam-affected people are indeed forced to migrate thus to radically change their place of residence and their life plans. Settlement of dam-affected families must be done relatively near the construction site. A new location implies administrative, political, logistical, and cultural changes, adding to all the adjustments due to migration.

When settling into a new administrative framework (municipality and/or state of the federation), families need to adapt to a new living environment and potentially establish new relations with the institutions, that is to say the public authorities, governments, associations, companies, etc. The institutional setting is important in the lives of families because it provides stability and security. A change causes upheavals. More specifically, changing banks, unions, links with local politicians, etc., weighs heavily on the families' daily life and affects their well-being and their future. Making contact and establishing a new relationship of trust with the various stakeholders are long-term processes that will impact the way families cope.

For example, families need to adapt to a new healthcare environment and familiarize themselves with new procedures to see doctors, go to the dentist, visit a clinic, etc. The examples are even more striking for the populations who settle in rural areas because this requires increased mobility and longer travel times. The victims of the Machadinho Dam, between Rio Grande do Sul and Santa Catarina, provide a good example. Some affected families had to migrate from the south to the north, thus changing their state of residence. The shock in terms of relations with health institutions was important because the southern state offers different services, sometimes considered more comprehensive. For instance, a mobile dentist service, traveling by bus, exists in rural areas, unlike in Santa Catarina, and the number of dispensaries in some municipalities of Santa Catarina is lower than in Rio Grande do Sul. Families in the south complained particularly about queues and the quality of the service offered in Santa Catarina.

Integration in a new territory can be facilitated to a different extent depending on the terms of the families' settlement. Families who settle individually do not benefit from the same support as those who arrive in groups. Sometimes, there is a collective settlement arrangement, mainly in rural areas, which facilitates the integration of families. The collective rural *reassentamento* (resettlement) gives families the same experience for their installation in their new place of residence. They can support each other and move forward together to build their new living environment. They may also benefit from basic infrastructures sometimes provided by this type of arrangement: a community center, a church, a school, telephone and electrical lines, etc. The families of collective *reassentamento* integrate more easily, both in the new community that they compose, but also in the new local rural and administrative areas of the municipality. A group of families going through the same experience can help each other and learn together about their new living place. It is however important to acknowledge that, despite benefiting from the same conditions as others, some families still struggle to settle into the new environment.

Let us look at a paradoxical example that shines another light on the issue of integration and exclusion. A group of families having received late collective compensation from the company responsible for the Machadinho Dam (MAESA) relocated to another state (Santa Catarina), several dozens of kms from the urban center of Curitibanos. They did not benefit from the same infrastructures as the other *reassentamento* established in the region. Due to various historical reasons, conflicts, and negotiations, the initial group of families settling in this new space had drastically reduced in the first few months. The living conditions being particularly difficult (no house built, very little help from institutions, isolated place), many families chose to leave and gave up their compensation. Nevertheless, despite the complexity of the living environment, some families decided to stay and unite to settle successfully. During the research in this *reassentamento* (field research in 2007), we noted a real unity and a very supportive environment among the families. They helped each other a lot during the first few months to build houses and plant the first crops. They then continued to meet at the end of each day to build together, with their own resources, a church and a community center. This collective action is also a way of strengthening the unity of the group and establishing roots in the territory.

Territorial exclusion affects rural communities present before the arrival of the dam. Some areas see their inhabitants leave because of the rising waters of the reservoir. Rural communities see their population drop and must adapt to a new context, that is to say a dissolution of interrelations between families and neighbors because of departures. Smaller communities have limited infrastructure, especially for businesses or community services, such as priests in churches. The rural community of São Francisco, in the municipality of Barracão/RS (Rio Grande do Sul), lost more than half of its inhabitants during the construction of the Machadinho Dam. At the same time, the loss was accompanied by the closure of businesses, a church, and a primary school. Public services were reduced due to a lack of families. A whole section of community life can disappear after the migration of families affected by a dam, and their original living areas become isolated.

Another type of exclusion concerns dam-affected families who migrate to a new living space but who are unable to settle permanently. We met families of "cabocla" origin who, in a collective rural *reassentamento*, did not manage to have an agricultural production and to feel comfortable in their new home. Excluded families who have lost their point of reference have difficulties in finding productive work, sometimes due to the necessity to adopt new farming practices. These families miss their former living environment (from a physical and psychological point of view). They suffer from the absence of parents and family, friends and neighbors, but also from the loss of their landscape, including the watercourse, which can be a fundamental element for their well-being and life structure. These families are however installed in *reassentamentos* where others are successful in their new life. Territorial exclusion can thus be expressed individually or collectively depending on the situation. The adaptation of families to the new environment affects their integration and their participation in the development of the new territory.

In his definition of territory, G. Di Meo speaks of ideological and political appropriation of space. The stakeholders who impact directly or indirectly on the territory are referred to as spatial actors. They influence and build the territory through their actions or their representations of the future territory. According to V. Noseda and J-B. Racine (Noseda and Racine 2001), the main geographical actors are: the individuals (including their family or their household), a more or less formal group, the company, the local community, and the state. Everyone may have a different field of action and a different strength, but all have a certain impact on the space and participate in the territorialization. These actors work according to their socio-spatial practices, their objectives, and their interests by setting up a strategy, based on domination, alliance, or cooperation. The spatial actors build the territory over time, and in this case, it is the period of the construction of a dam.

The construction of a hydroelectric dam implies the upheaval of the actors present in the area and thus the transformation of the territory. While new stakeholders appear, often powerful (economically, politically, in the decision-making domain, etc.), others disappear or see their fields of action and responsibilities diminish. This results in a power redistribution, especially economic and political powers, and a reconstruction of the territorial development with the arrival of a new set of actors in a specific area, i.e., the site of construction and the area of influence around the hydroelectric plant.

Dams are usually constructed in rural areas, which are dominated by agricultural stakeholders, mainly small farmers. They rarely get built in areas of large land holdings (fazendas) as powerful and influential owners tend to have more land on the plateaus than in the valleys (Monbeig 1952) and also have the means to prevent their construction. Small farmers therefore represent the largest group of stakeholders in rural areas before the arrival of a dam, despite the recent change in the profile of rural areas with an increase in non-agricultural populations (Balsadi 2011). For our field research, rural unions' headquarters have often been key locations, starting points for identifying key players in the local territory. Small farmers do not have a strong economic domination, but by their number and their networks, they are the main actors of the territorialization. They are accompanied by traditional government services, whether local communities or federal state services. Public services are still major providers of jobs in rural Brazil in all sectors: administration, education, health, security, etc. With the establishment of dams, rural areas are in turmoil, especially through the introduction of new players (tourism and recreation, economic center, environment, new inhabitants, etc.).

A few months before the start of construction, the set of actors changes completely with the arrival of the dam construction company and its unique economic strength and equally impressive symbolic power. The company comes to an area where the most powerful stakeholders are of less importance and quickly takes over by deciding many elements that mark the territory. In addition to building a dam and a plant, it introduces various other infrastructure and service elements. For example, companies may decide to asphalt roads, reinforce police services, build churches, etc. They model the territory to their liking, and the other stakeholders do not have enough weight to discuss and oppose their decisions. Private entities take power despite their governance responsibilities being “unclear, non-transparent, and conflicting” (Boanada et al. 2016). One example among many, the CHESF Company that built the Sobradinho Dam in the state of Bahia in the 1970s built villages to accommodate the employees responsible for the construction of the structure. Even after fulfilling their original objective, these villages continue to exist years later as many employees chose to stay on site after completion of the dam. Today, Sobradinho is a small town of 20,000 inhabitants (IBGE 2010) whose origin is due to the construction of the dam about 40 years ago.

The construction of a dam is often marked by a questioning of the federal government’s involvement, or at least of its representatives. Good part of the dam construction is the result of a global energy policy, and the establishment of a dam is, theoretically, of benefit to the entire national electricity sector. The division of responsibilities between the federal state and the construction company is sometimes blurred because the financing may come from national banks. There is a confusion between private and public spheres. The state company and its private partners may argue over issues, responsibilities, and obligations with respect to public policies or activities related to the dam development (Boanada et al. 2016). The confusion of roles and responsibilities is complex and intermixed with other issues, making it hard for the local population to determine who to refer to in case of specific problems. For example, during the construction of the Belo Monte Dam, the presence and influence of FUNAI in Altamira have greatly diminished despite the presence of indigenous people in the impact zone of the dam. The number of officials of this public institution that protects the native indigenous populations dropped from 60 in 2011 to 23 (officially) in 2015, but most probably only 14 because of early departures (Indicadores de Belo Monte 2015 and G1 2015). FUNAI was therefore greatly weakened during the construction of the dam, a time during which they were most needed for the protection of indigenous people in the region, demonstrating the influence of the private sphere on the public sector. The local population, and especially indigenous people, are worried about this development and demonstrated in 2017 against the downsizing of FUNAI. In addition to a national policy of withdrawal from this foundation, the influence of Norte Energia has reduced the action of the institution in the region. The company can now deal with the indigenous families in its own way, deciding their future and therefore their territory in the region.

New stakeholders, beyond the construction company, also arrive in the dam area, such as other companies involved in construction, associations, or NGOs and public control entities. A whole new set of spatial actors settle in the area and remodel the territorial organization of the region by acting directly in the local space and very often ignoring the local population. These new stakeholders often replace previous local actors, and the balance of power is upset. The new actors act in the space according to their own interests and tend to settle in the area for a short period of time, namely the duration of construction. They seize the decision-making powers and the spheres of influence. Even associations and non-governmental organizations, that act to protect the population or the environment, mark the space and participate in the construction of the territory. The example of the social movements in defense of the victims of dams illustrates this point. Whether well-established and nation-wide (such as the Movement of People Affected by Dams (MAB)), or new and local only, these organizations mark the territory by their grip on the dam-affected populations. The social movements organize meetings, demonstrations, and protests to claim land, better compensation, more negotiations, etc. By their protest actions, they create a territory of opposition against the dam company. This occurs very often during construction and sometimes even before the start of work. Before the beginning of Foz do Chapecó, the MAB installed a camp of local families who took turns to occupy the land marked out for construction, thus imposing their presence from the outset to negotiate compensation.

The stakeholders that were previously in a position of strength are left behind with regard to governance and territorial development. Local public entities are often subjected to decisions from above, either from a broader political power or from an influential economic entity. Spatial planning, which in Brazil is largely the responsibility of local communities, passes into the hands of the company building the dam and its associates. Territorial development takes a functional turn responding to the needs of economic actors. It is possible to question the real place of public power during the construction of dams because they sometimes seem very far behind. The city of Altamira is an enlightening example of the loss of power of the main local stakeholder, namely the town hall. To compensate for the negative effects of the Belo Monte Dam, Norte Energia undertook to set up a sewerage and water treatment system in the city. During interviews with town officials responsible for planning and public works, they explained not being aware of the project of the sanitation system and to have no information on the schedule or on the location or terms of operation. City officials did not know what was happening in their own city regarding the installation of a new public service for the population. The loss of power of a local actor is the consequence of the arrival of a new actor who takes over and decides the territorial future around the dam.

Similarly, local populations participate relatively little and in an illusory way in the decision-making process regarding the future of their living environment. The Brazilian environmental licensing system provides for public hearings, but these are conducted in such a way that they present a project already set up and approved by the company and the public authorities. Thus, the local population has no say in the official course of construction of a dam. For the company, the public hearings are controlled and organized and allow above all to legitimize their action publicly. In this way, there is a clear lack of space for dialogue and debate between civil society and the private sector (Locatelli 2014) (Cf. Fig. 1.5).

All the stakeholders, new and old, have the future of the region in their hands and decide the territorial development according to their desires and powers. Regions composed of various territories see their functions modified according to

Fig. 1.5 Meeting between future atingidos and federal government during 30 min in Rio dos Indios/RS. Leturcq (2005)

the construction of the dam and the many actors and spheres of influence on site. For example, after the wave of dam constructions on the Uruguay River, the city of Florianópolis, capital of the Santa Catarina state, became a logistical, administrative, political, and decision-making center for the Brazilian electricity sector in the southern region. This city brings together various key players in the sector.

The economies of the areas hosting the dams are modified by the arrival of the dams, in particular the agricultural sector and all that it incorporates: the loss of useful agricultural land, the loss of productive agents in this sector because of migrations, a change in the profile of the local workers, and a lack of economic interest due to the new dominant sector.

The territories around the dams are constantly redefined with several new or modified features: economies, stakeholders, institutions, etc. The local populations are in the midst of the transformations and have a reduced space of participation regarding the evolution of the territories. The construction of a dam is always a territorial element that becomes central, and it is often around it that the local and regional space gets restructured.

References

Akinaga P, Namba M, Cunha I, Vinueza G, Fontes GL, Negro JRA (2010) Paisagem recriada: Projeto de descaracterização das barragens da mina da cachoeira. Rev. Lab Verde 1:117–133

ANEEL—Agência Nacional de Energia Elétrica (2008) Atlas de energia elétrica do Brasil. ANEEL, Brasília, 3 ed, p 236

Balsadi OV (2011) Mudanças no meio rural e desafios para o desenvolvimento sustentável. São Paulo Perspec 15:155–165

Barrow C (1988) The Impact of hydroelectric development on the amazonian environment: with particular reference to the Tucuruí project. J Biogeogr 15:67–78

Benetta CJD (2002) Itaipu: um megaprojeto. In: Dos Santos SC, Nacke A (org) Memória do Setor Elétrico na Região Sul. UFSC, Florianópolis, pp 130–145

Bermann C (2012) O projeto da usina hidrelétrica Belo Monte: a autocracia energética como paradigma. Novos Cadernos NAEA 15:5–23

Berque A (1995) Espace, milieu, paysage, environnement. In: Bailly A, Ferras R, Pumain D (org) Encyclopédie de Géographie. Economica, Paris, pp 349–366

BIG—ANEEL (2017) Banco de Informações de Geração—ANEEL. http://www2.aneel.gov.br/aplicacoes/capacidadebrasil/capacidadebrasil.cfm

BNDES—Banco Nacional de Desenvolvimento Econômico e Social (2004) BNDES alavanca com R$ 10 bilhões projetos de geração de energia elétrica. BNDES, Rio de Janeiro. https://www.bndes.gov.br/wps/portal/site/home/imprensa/noticias/conteudo/20040623_not834

Boanada V, Leturcq G, Grisotti M (2016) Acceptabilité sociale et place de la population lors de la construction du barrage de Belo Monte (Brésil). Éthique publique 18:1. http://journals.openedition.org/ethiquepublique/2540

Brzezinski MLNL (2014) Para entender Belo Monte: a história de uma usina hidrelétrica contada pelo jornal O Estado de São Paulo. Cadernos IHU 47:20–37

Broeckelman M (1979) A barragem de Sobradinho na Bahia/Brasil: desenvolvimento para quem?. Reforma Agrária, v. 9–5

Caubet CG (1991) As Grandes Manobras de Itaipú. Energia, Diplomacia e Direito na Bacia do Prata. Acadêmica, São Paulo, p 385

Caubet CG (2004) A Água, a Lei, a Política e o Meio Ambiente. Juruá, Curitiba, p 206

Cernea MM (2000) Risks and reconstruction: experiences of resettlers and refugees. World Bank Publications, Washington DC, p 487

Cintra IHA, Flexa CE, Silva MB, Araújo MVLF, Silva KCA (2013) A pesca no reservatório da usina hidrelétrica de Tucuruí, região Amazônica, Brasil: aspectos biológicos, sociais, econômicos e ambientais. Acta Fish Aquat Res 1:57–78

Comissão Pastoral da Terra de Rondônia (2011) Prostituição e pedofilia nas usinas do Madeira e na fronteira da Bolívia. http://cptrondonia.blogspot.com.br/2011/12/prostituicao-e-pedofilia-nas-usinas-do.html

Companhia Hidrelétrica do São Francisco-Chesf (2017) Descrição do Aproveitamento de Sobradinho. http://www.chesf.gov.br

Corrêa ML (2003) O setor de energia elétrica e a constituição do Estado no Brasil: o Conselho Nacional de Águas e Energia Elétrica (1939–1954). PhD Thesis, Universidade Federal Fluminense, Niterói

Couto RCS (2002) Malária: o custo social da Hidrelétrica de Tucuruí-PA, Brasil. In: Couto RCS, Castro EMR, Marin RA (eds) Saúde, trabalho e meio ambiente: políticas públicas na Amazônia. Universidade Federal do Pará, Belém, pp 107–121

CTG—China Three Gorges Corporation (2017) The company. http://ctgbr.com.br/en/the-company

Daou AM (1988) Políticas de Estado e Organização Social Camponesa: a barragem de sobradinho. Master Dissertation, UFRJ, Rio de Janeiro

Destaque Amazônia (1985) Parakanã: a luta por nova demarcação de sua Reserva

Di Méo G (1998) Géographie sociale et territoire. Nathan, Paris, p 320
Duqué G (1984) A experiência de Sobradinho. Cadernos do Ceas, Salvador, 91:30–38
Eletrobrás, Eletronorte, MME (2009) Relatório de Impacto Ambiental—RIMA. Aproveitamento Hidrelétrico de Belo Monte. http://eletrobras.com/pt/AreasdeAtuacao/geracao/belo_monte/Rima-AHE-Belo-Monte.pdf
Fearnside PM (1999) Social impacts of Brazil's Tucuruí Dam. Environ Manag 24(4):483–495
Fearnside PM (2008) Hidrelétricas como "fábricas de metano": O papel dos reservatórios em áreas de floresta tropical na emissão de gases de efeito estufa. Oecologia Brasiliensis 12(1):100–115
Fearnside PM (2011) Hidrelétricas amazônicas como emissoras de gases de efeito estufa. Proposta 35(122):24–28
Fearnside PM (2014a) Análisis de los Principales Proyectos Hidro-Energéticos en la Región Amazónica. Derecho, Ambiente y Recursos Naturales, p 55
Fearnside PM (2014b) Impacts of Brazil's Madeira River dams: unlearned lessons for hydroelectric development in Amazonia. Environ Sci Policy 38:164–172
Fearnside PM (2015) Impactos Sociais da Barragem de Tucuruí. In: Hidrelétricas na Amazônia: Impactos Ambientais e Sociais na Tomada de Decisões sobre Grandes Obras. Vol. 1. Editora do Instituto Nacional de Pesquisas da Amazônia (INPA), Manaus, p 296
Fearnside PM (2016a) Greenhouse gas emissions from hydroelectric dams in tropical forests. In: Lehr J, Keeley J (eds) Alternative energy and shale gas encyclopedia. Wiley, New York, p 912
Fearnside PM (2016b) Brazilian politics threaten environmental policies. Science 353:746–748
Fearnside PM (2017a) Belo Monte: actors and arguments in the struggle over Brazil's most controversial Amazonian dam. Erde 148(1):14–26
Fearnside PM (2017b) Brazil's Belo Monte Dam: lessons of an Amazonian resource struggle. Erde 148(2–3):167–184
Ferreira JZ (2008) Sustentabilidade e participação social antemega-dams na amazônia. O caso PPDJUS a Jusante da UHE Tucuruí e o Setor Elétrico Nacional. IV Encontro da Associação Nacional de Pós-Graduação e Pesquisa em Ambiente e Sociedade. http://www.anppas.org.br/encontro4/cd/ARQUIVOS/GT4-462-282-20080515234447.pdf
Filho AOS (org) (2005) Tenotã-Mõ: alertas sobre as consequências dos projetos hidrelétricos no Rio Xingu. São Paulo, International Rivers Network, p 344
Fleury LC (2013) Conflito ambiental e cosmopolíticas na amazônia brasileira: a construção da Usina hidrelétrica de Belo Monte em perspectiva. PhD Thesis, Universidade Federal do Rio Grande do Sul, Porto Alegre
Fundação Getulio Vargas (2009) Verbete tematico CHESF. http://www.fgv.br/cpdoc/acervo/dicionarios/verbete-tematico/companhia-hidro-eletrica-do-sao-francisco-chesf
G1 (2017) Índios protestam contra medida que reduz número de servidores da Funai. https://g1.globo.com/pa/para/noticia/indios-protestam-contra-medida-que-reduz-numero-de-servidores-da-funai.ghtml
Germani GI (2003) Expropriados terra e água: o conflito de Itaipu, 2e edn. EDUFBA-ULBRA, Salvador, p 266
Glass V (2013) Adolescente é resgatada de prostíbulo em Belo Monte. Reporter Brasil.http://reporterbrasil.org.br/2013/02/adolescente-e-resgatada-de-prostibulo-em-belo-monte/
Goldemberg J (2008) O Banco Mundial e as hidrelétricas. O Estado de S. Paulo. http://www.eln.gov.br/opencms/opencms/modulos/noticia/noticia_0254.html?uri=/index.html
Guimaraes JRD (2001) Les processus de méthylation du mercure en milieu amazonien. In: Carmouze J-P, Lucotte M, Boudou A (eds). Le mercure en Amazonie: rôle de l'homme et de l'environnement, risques sanitaires. IRD, Paris
Gomes ACS, Abarca CDG, Faria EAST, Fernandes HHO (2006) O Setor Elétrico. Livro Setorial do Banco Nacional de Desenvolvimento Econômico e Social (BNDES). http://www.bndes.gov.br/conhecimento/livro_setorial/setorial14.pdf

IBGE—Instituto Brasileiro de Geografia e Estatística (2010a) Censo Demográfico 1970. IBGE, Rio de Janeiro
IBGE—Instituto Brasileiro de Geografia e Estatística (2010b) Censo Demográfico 1980. IBGE, Rio de Janeiro
IBGE—Instituto Brasileiro de Geografia e Estatística (2010c) Censo Demográfico 2010. IBGE, Rio de Janeiro
IHA—International Hydropower Association (2017) Hydropower status report. IHA, London
Indicadores de Belo Monte (2015) Número de servidores da Funai (Geral, Diretoria de Proteção Territorial e Altamira). http://indicadoresdebelomonte.eco.br/metrics/337
International Network Rivers (2008) Social and Environmental Standards for Large Dams. IRN. www.internationalrivers.org/node/3959
IPEA—Instituto de Pesquisa Economica Aplicada (2014) Capacidade instalada - geração de energia elétrica MW. Ministério de Minas e Energia, Balanço Energético Nacional (MME)
IPEA—Instituto de Pesquisa Economica Aplicada (2017) Atlas da Violência 2017. Rio de Janeiro. http://www.ipea.gov.br/atlasviolencia/download/2/2017
Leite L (2015) Belo Monte: os filhos da barragem. Instituto Socioambiental-ISA. http://www.socioambiental.org/pt-br/noticias-socioambientais/belo-monte-a-violencia-na-compensacao
Locatelli C (2014) Communicação e barragens: O poder da communicação das organizações e da mídia na implantação de hidrelétricas. Florianópolis, Insular, p 360
Louzada AF (2009) Conflitos por múltiplos usos da água no reservatório da UHE de Tucuruí. Master Dissertation, Universidade Federal do Pará, Belém
Magalhães SB (2005) Política e sociedade na construção de efeitos das grandes barragens: o caso Tucuruí. In: Filho AOS (ed) (2005) Tenotã-mõ. International Rivers Network, São Paulo, p 344
Magalhães SB (2007) Lamento e dor. Uma análise sócio antropológica do deslocamento compulsório provocado pela construção de barragens. PhD Thesis, Universidade Federal do Pará, Université Paris, Belém
Magalhães SB, da Cunha MC (2017) A Expulsão de Ribeirinhos em Belo Monte: Relatório da SBPC. Sociedade Brasileira para o Progresso da Ciência, São Paulo, p 448
Magalhães SB, Hernandez FM (org) (2009) Painel de especialistas: Análise Crítica do Estudo de Impacto Ambiental do Aproveitamento Hidrelétrico de Belo Monte. Belém
Martins DP (2008) Aménagements hydroélectriques et impacts sur la dynamique des flux d'eau et de sédiments. PhD Thesis, Université Lumière Lyon 2, Lyon
McCully P (2001) Silenced rivers: the ecology and politics of large dams. Zed Books, New York, p 359
Mello FM (2011) Um Século de Obras contra as Secas. In: Mello FM (ed) A história das barragens no Brasil, Séculos XIX, XX e XXI: cinquenta anos do Comitê Brasileiro de Barragens. CBDB, Rio de Janeiro, p 524
Meritet S (2004) Vers une nouvelle crise de l'industrie électrique au Brésil. La revue de l'énergie 555:6
Monbeig P (1952) Pionniers et planteurs de l'État de São Paulo. Armand Colin, Paris, p 376
Moreira R (2006) Para que o EIA/RIMA quase vinte anos depois? In: Verdum R, Medeiros RMV (org) RIMA: Relatório de impacto ambiental, 5e edn. Universidade Federal do Rio Grande do Sul, Porto Alegre, pp 13–18
Mougeot L (1988) Reinstação de populações na Amazônia Brasileira. O Caso de Tucuruí. In: Aubertin C (ed) Fronteira. UNB, Brasília
Noseda V, Racine JB (2001) Acteurs et agents, points de vue géographiques au sein des sciences sociales. Revue européenne des sciences sociales XXXIX-121
Nogueira ICG (2010) Segregação socioespacial urbana no entorno de hidrelétrica: produção do espaço em Tucuruí-PA. Master Dissertation, Universidade da Amazônia, Belém
Porvari P (1995) Mercury levels of fish in Tucuruí hydroelectric reservoir and in River Mojú in Amazonia, in the state of Pará, Brazil. Sci Total Environ 175(2):109–117
Queiroz R (2013) Setor elétrico brasileiro: uma história de reformas. Energia elétrica. https://infopetro.wordpress.com/2013/09/09/setor-eletrico-brasileiro-uma-historia-de-reformas

Rossato A (2013) Trabalho escravo nas hidrelétricas financiadas pelo BNDES. Comissão Pastoral da Terra Nordeste II. http://www.cptne2.org.br/index.php/publicacoes/noticias/trabalho-escravo/2499-trabalho-escravo-nas-hidreletricas-financiadas-pelo-bndes

Sakamoto L (2011) IBAMA, Belo Monte e um rolo compressor. UOL Noticias. Blog: blogdosakamoto.blogosfera.uol.com.br/2011/01/12/ IBAMA-belo-monte-e-um-rolo-compressor/

Santos SC, Nacke (org) A (2003) Hidrelétricas e povos indígenas. Florianópolis, Letras contemporaneas, p 178

Santos JML, Mello SCB (2014) Os significados de Belo Monte no discurso das mídias sociais. Revista CMC, 11(30):13–36

Schmitt JV (2008) Os atingidos por Itaipu: História e memória. Oeste do Paraná, decadas de 1970 a 2000. Master Dissertation, Universidade Estadual do Oeste do Paraná, Marechal Cândido Rondon

Sigaud L (1988) Efeitos sociais de grandes projetos hidrelétricos: as barragens de Sobradinho e Machadinho. In: Rosa LP (ed) Impactos de grandes projetos hidrelétricos e nucleares: aspectos econômicos, tecnologicos, ambientais e sociais. Marco Zero, São Paulo, pp 83–165

Sigaud L (1992) O efeito das tecnologias sobre as comunidades rurais: o caso das grandes barragens. RBCS, São Paulo, nº 18, ano 7

Silva EM, Germani GI (2009) Os que não foram e os que voltaram: a rejeição ao PEC/Serra do Ramalho como resistência camponesa a sobradinho. IV Simpósio Internaciolnal de Geografia Agrária, Rio de Janeiro

Terras Indigenas no Brasil (2002) Reserva Indígena Avá Guarani/Ocoí. https://terrasindigenas.org.br/es/terras-indigenas/3606

Uderman S, Cavalcante LR (2011) O papel do BNDES no financiamento do setor de energia elétrica no Brasil. Parcerias Estratégicas 16(33):27–280

Vainer CB (2004) Águas para a vida, não para a morte. Notas para uma história do movimento de atingidos por barragens no Brasil. In: Acselrad H, Herculano S, Pádua JA (Org) Justiça Ambiental e Cidadania, vol 1. Relume Dumará, Rio de Janeiro, pp 185–216

Vainer CB (2007) Recursos hidráulicos: questões sociais e ambientais. Revista Estudos Avançados 59

WCD (1999) Estudos de Caso da Comissão Mundial de Barragens (CMB), Usina Hidrelétrica de Tucuruí (Brasil) Relatório Final da fase de escopo. http://www.ib.usp.br/limnologia/textos/Usina%20Hidrel%E9trica%20de%20Tucuru%ED%20(Brasil)%20RELAT%D3RIO%20FINAL%20DA%20FASE%20DE%20ESCOPO%20AGOSTO%20DE%201999.pdf

WCD (2000) Dams and development: a new framework for decision-making. World Commission on Dams

Wolkmer MF, Pimmel NF (2013) Política Nacional de Recursos Hídricos: governança da água e cidadania ambiental. Seqüência: Estudos Jurídicos e Políticos 34:165–198

Zhouri A, Oliveira R (2008) Développement et violence sociale dans le Brésil rural: le cas des barrages hydroélectriques. Ecologie & Politique 35(1):133–145

Chapter 2
Human and Social Consequences

Abstract In this chapter, the focus which guides our analysis is geographical. The first aspect concerns the migration which is caused by the dams. There are many complex migrations associated with the construction of a dam, which take place at different phases and for several population profiles. These migrations influence the livelihood of the affected families who must move and adapt to a new environment, under specific circumstances and conditions. Economic, social, and structural factors affect the degree of the families' adaptation. Adaptation also depends on the pre-migration period, a key phase for families when they have access to compensation, when they must negotiate, and when they face various types of violence. Finally, after looking at these challenges and the peculiarity of the phenomenon, we will analyze the identity of the affected population which develops through the challenges and struggles. This chapter presents a wide range of social repercussions associated with dams in Brazil.

2.1 Migrations Aspect

2.1.1 Pre-migrations

Population migrations are one of the consequences of dam construction in Brazil. Migrations have accompanied the country's development for hundreds of years, but hydroelectric plants are a source of additional migrations of another nature and of variable amplitude.

Brazilian history is marked by the displacement of populations. Colonized from the sixteenth century, Brazil's first inhabitants were first hunted and evicted by settlers in coastal areas. Brazilian Indians were the first groups forced to change their homes and places of life. A few decades later, the Portuguese settlers set up a strategy of occupation of the interior lands of the continent, after having occupied thousands of kms of coastline. Subsequently, bandeirantes, mercenaries, organized violent expeditions in the Brazilian countryside, following specific routes. At each stage of their journey, they left some families to occupy the space and to

G. Leturcq, *Dams in Brazil*, SpringerBriefs in Latin American Studies,
https://doi.org/10.1007/978-3-319-94628-3_2

appropriate the lands. In just a few years, the Portuguese claimed to have populated many regions and claimed distant territories, thus affirming their strategic positions against the Spanish settlers. They covered the territory, expanding borders and controlling it with a limited number of men (Droulers 2001).

Migrations of populations continued to accompany the construction of the country during the various stages of clustered development. As the country grew during the economic cycles of sugar, coffee, and rubber, many people moved to centralized locations (Droulers 2001). For example, from the end of the nineteenth century, Brazil became the world's largest rubber producer and the entire national economy benefited from this position (40% of national exports (Dean 1989)). At the same time, people started leaving the poor regions of the Nordeste to go to the Amazon in search of jobs and economic opportunities (Leturcq 2016). Furtado (1970) estimates that 200,000 migrants moved to the Amazon in the last decade of the nineteenth century. The total number of migrants for the rubber boom (1850–1915) is estimated at 350,000 by Benchimol (1977). The population flow was important, both in terms of quantity and influence on the territories and their future, because migrants chose to settle in the Amazon.

The late nineteenth and early twentieth centuries were also a period of colonization, wanted and organized by the Brazilian government. Families coming first from Germany, Italy, Portugal, Eastern Europe, then Japan settled mainly in the south and southeast of the country. The idea was to occupy and develop large agricultural areas, including the Rio Grande do Sul countryside (Roche 1959). Between 1884 and 1933, almost 4 million migrants came to settle in Brazil (IBGE 2007).

The second half of the twentieth century is also characterized by a large-scale migration: a massive rural exodus that fed a rapid urbanization. In 50 years (1960–2010), Brazil went from a rural population of 45% to about 15% (IBGE 2010), causing a spectacular urban concentration. According to Maia and Buainain (2015), this phenomenon persists despite a slowdown since the 2000s. In Brazil, "since 1940, every decade, over one-third of the population who lived in rural areas at the beginning of the period moved to the cities" (Abramovay 1998). Rural areas are neglected but not totally emptied, for economic reasons and due to the concentration of land holdings. On the other hand, economically attractive urban centers receive often young, male, and poor populations (Maia and Buainain 2015) coming from all areas of the country. In the 1950s, two-thirds of Brazilians lived in the countryside, while in the late 1990s three quarters were installed in the cities (Théry 2005). The concentration of population is therefore very important with more than 84% of the population living in urban areas (IBGE 2010).

The first dam constructions started after 1960 and therefore took place in the context of the rural exodus of the twentieth century. Dam location sites are not necessarily different from the rest of the country, but they present different population dynamics. For example, people have been living in the area around the Uruguay River, in the south of Brazil, for more than 100 years, after the arrival of European immigrants. As the last occupied spaces in the southern region, the banks of the Uruguay River saw families leave the area from the second half of the twentieth century due to the scarcity of available land. J. Roche (Roche 1954) explains that settlers, who tended to keep the same birth rate as in their native country, quickly found themselves running out of land for their descendants who had to migrate to "virgin" lands in the north (Santa Catarina, Paraná, Mato Grosso, Amazonia).

On the contrary, population flows linked to dam construction in the north of the country were different, with an expulsion of native indigenous populations and the arrival of people coming from the south or Nordeste. According to INCRA's plans, 75% of migrants were meant to come from the Nordeste and the other 25% from the south (Moran 1981). The works of Becker (1982) and Moran describe the political land use plans and the population movements that accompany them. The military power of the 1960s wished to exploit and occupy the north of the country. To this end, they encouraged a migration of settlers to occupy agricultural lands made accessible by the construction of the Trans-Amazonian Highway. Xavier Arnauld de Sartre speaks of "the perpetuation of geophagic-type space occupation models" (Arnauld de Sartre 2006). A development policy, the "Amazon operation," was conceived in the 1960s and implemented the following decade. It began with the creation of the National Integration Plan in 1970 for the exploitation and integration of natural resources in the Brazilian Amazon (Moran 1981). In this context, newly built dams, such as Tucuruí (1984) and Belo Monte (2016), reinforced migratory flows toward these areas.

In these settings, the arrival of a dam upsets, at least in the medium term, the rationale for local and regional migratory flows, creating violent distortions that are difficult to control.

The population migrations that accompany the construction of hydroelectric dams can be grouped into two categories: on the one hand, the direct victims of the dams, which are the populations forced to migrate and to change their place of residence because it has been destroyed by the construction; on the other hand, the populations that migrate to accompany the construction of the dam looking for opportunities or to change their life. Both types of migration follow the information available to the individuals and families. Communication and access to information are fundamental to understanding migrations.

The construction of a dam takes place over an extended period of time. It may take several decades between the creation of the first plans and the completion of the work. It is therefore not uncommon to meet families in rural areas who have heard about a dam development on a nearby river but who still know nothing about its actual construction. In the time between the engineers' plans and their execution, there is a lengthy decision-making process which greatly depends on the politico-economic context. Each dam is the result of a decision-making process that is often far removed from the local realities, and a certain time is required for the decisions to reach the local inhabitants. L. Sigaud describes "the incredulity of the peasants" (Sigaud 1995) and the lack of understanding of the events. During one of our field trips, we met a family living less than 5 kms from the building site of the

Foz do Chapecó Dam on the Uruguay River who was not aware of its construction, because they did not have a direct view of the site due to a meander on the river.

Communication is essential to help dam-affected families to better understand what's happening and to prepare themselves for the future consequences. The government, often discreet and supportive of the construction of hydroelectric plants, adopts a positive communication, favorable to development, as confirmed by Locatelli in his book "Comunicação e Barragens" (2014). The government always keeps a certain distance and sometimes communicates as little as possible, or with restraint, so as not to trigger the media machine, especially on some controversial projects. Government representatives are, however, often present at the launch of the construction or the inauguration of the plant, such as President Dilma Rousseff during the inauguration of the Belo Monte Dam on May 5, 2016.

The communication from the representatives of the population remains very weak in terms of diffusion and weight compared to that of the company responsible for the construction of the dam. Social movements must diversify their messages and act on different levels. As Maria da Glória Gohn explains (Gohn 2003), "the visibility of the actions is present in different spaces: in small decentralized groups that choose their own mode of participation; in public and collective spaces; and in global spaces, the media, and therefore is part of our everyday experiences." Unlike companies, social movements, trade unions, and opponents of dams have a few means and resources to put in place an effective communication plan, but they manage to get their messages across with varying success depending on the situation.

The companies responsible for the construction of the dams use all their economic weight to influence the local population by promoting as much as possible the supposed benefits of the dam. Locatelli explains (2014), using the example of the company Foz de Chapecó Energia (FCE), that the companies put in place a very effective communication strategy. It is marked by an absence of debate and by direct communication in small groups (meetings), with the population and its representatives, to fully exercise its economic, political, and symbolic powers. The companies responsible for the dams thus assume a prominent place in spreading the image of the dam and make local people accept the benefits of construction. In the case of the FCE Company, we have identified nine different methods and means of communication to disseminate messages (Leturcq 2010). Businesses are increasingly taking on the role of communicators previously assumed by public institutions. During this process of role substitution, the emphasis on public debates always decreases, while content and relevance diminish, so various questions are ignored or forgotten (Locatelli 2014).

The Brazilian state only gives a general framework in economic, social, political, and environmental terms and leaves the daily activities and the general management of the dam construction in the hands of the private sector. According to C. Locatelli, the "deficit of democracy in terms of information is damaging for the populations" (Locatelli 2014). In its 2000 report, the World Commission on Dams emphasized the importance of "prior and informed consent" from both formal and informal representative organizations. Information monitoring is essential for the future of dam-affected families in terms of knowledge and decisions regarding compensation. Beforehand, the families must, however, realize the inevitability of the dam and its direct repercussions on their lives and territories.

A change of residence is a significant consequence for families who are victims of the dams. The families must leave their home and their place of life to move to a new, unknown, and distant area. The migration process can take place over several years (between 2 and 5 years), between the first assessments of the lands soon to be flooded and the actual move. Delay is an important criterion, which plays a big part in understanding the phenomenon and preparing for the inevitable change. As the duration is undetermined, the preparation is truncated and uncertain, making the situation particularly complex for families relying on agriculture and therefore following precise calendars to ensure good yields. During this period, feelings of uncertainty often arise, which can vary in intensity and scale depending on the family.

Preparation for the arrival of a dam begins with a need for key information on whether or not it will actually be constructed. There are many hydroelectric construction projects in the drawers, but certain conditions must be met for the projects to succeed. Local people are aware of the projects but do not always know the date of commencement of the works and have no way to follow the directives and decisions on the construction choices.

Some families curious to find out and to learn more about lasts events meet local social organizations. Churches, for example, can play an informational and organizational role. Families, with the help from institutions, can better prepare themselves and have a more detailed idea of what can happen.

Some families eagerly wait for a change of residence in the hope to start from scratch and begin a new life. They even envision how to take advantage of their new home or the compensation received to move forward and make a fresh start. Families met spoke of this situation as "winning the lottery" (Field Camp in 2007, Campos Novos/SC (Santa Catarina)). For some families, it is a question of getting out of a precarious situation in terms of housing or economic activity and trying to rebuild a life on new foundations thanks to the help and assistance that they can sometimes benefit from. We must never forget that the construction of a dam can indeed sometimes be seen as a positive event for families who see them as an opportunity.

On the contrary, other families may fear the future move. They may not be able to imagine or project themselves into the future for lack of information. We conducted interviews with people who refused to accept the reality or were determined to stay in their place of residence. The fear of the unknown can be strong, especially for the elderly who must break away from old roots in their territories. Cases of depression are often reported during the construction of dams in Brazil (in addition to cases met personally in Chapecó/SC and Altamira/PA (Pará), we can mention work with populations in the states of Paraiba and Minas Gerais), as with other dams in the world. Tangible and intangible losses are poorly accepted by many families who equate the situation with abandonment. Migrants have to leave behind a lot of memories, goods, and spatial references, which affects their morale. It is not uncommon, for example, to see cemeteries engulfed by the waters of a reservoir and thus to witness the loss of a place of remembrance for entire families and communities.

Finally, migration is often synonymous with separation between families and the breaking of existing social relations. The construction of a dam destroys links created between neighboring families in a community. There is never a plan to resettle an entire community in the same space to maintain existing links. Similarly, family members who are victims of the dam have little chance of being able to move near each other, unless they decide to organize themselves using individual financial compensation. The social bonds built on the territories disappear with the rise of the water, but new links are formed and some continue in spite of distance.

2.1.2 Types of Migrations

Migration studies deal with population movements, especially those that change their main place of residence, but they can also include changes in employment and in social relations (Simmons 2013). More often, they concern voluntary and thoughtful changes which respond to a certain logic, often economic. Economic migration corresponds to the idea of a change of residence. The construction of a dam is an economic event that motivates a displacement of population, but it also drives another kind of migration, constrained and forced by an event beyond the control of families. The forced migrations of dam-affected families are understudied. They affect a large group of families in Brazil because all hydroelectric dams result in home evictions. Finally, migrations occurring after completion of a dam, namely return migrations, also deserve attention.

Economic migrations are voluntary displacements of populations. The construction of a structure such as a hydroelectric plant creates jobs. Depending on the size of the dam, the quantity of jobs created can be considerable. The largest works, such as Belo Monte or Itaipu, require tens of thousands of workers. Dams are notorious projects which the government uses as a propaganda tool to promote the development they bring at all levels. Advertising is also orchestrated by the companies responsible for the dams. In their environmental impact reports, mainly RIMA, they insist on the creation of direct and indirect jobs. Locally, they justify the dam construction by creating jobs which, in rural areas that are sometimes not very dynamic, can be life-saving. It is almost the only justifying argument which is also used to mask the negative effects and other consequences of the work.

The RIMA of the Teles Pires hydroelectric plant states how many jobs were to be created. Figures were advanced (10,000 direct jobs and 20,000 indirect jobs), and the number of jobs was described as "significant" (EPE 2010) and offering "opportunities" to earn income. Job advertising also exists in newspaper articles and Internet headlines. The job creation argument accompanies the wider argument of economic growth promised for the region and the overall development of Brazil. The dam publicity has effects on the population, especially the poor. Families anticipate the beginning of the works and migrate, often in precarious situations. They come to live around the site looking for an opportunity and/or a job. These disordered and spontaneous migrations are often neglected by companies but are almost systematic. Hydroelectric plants are attractive prospects, and the publicity reinforces this attractiveness. These migrations, which could be anticipated and controlled, complicate the local territorial situation and fuel disorganization. Observed prostitution, violence, and drugs around dam construction sites demonstrate the lack of control on the part of the company.

The demographic growth of the municipality of Tucuruí as a consequence of the dam construction is edifying. According to the 1970 census (IBGE), before the arrival of the dam, the municipality had a little less than 10,000 inhabitants, 55% of them born locally. By 1980, the population had been multiplied by six and 77.3% of the inhabitants were born outside Tucuruí. The population continued to grow a further 33% in the next 11 years. In 1991, more than 67% of the inhabitants were born in the municipality, due to the reproduction of immigrants who remained in the region. The mass migratory movement from the outside coincides with the period of construction of the dam, and Tucuruí also welcomed landless populations coming to occupy new farmland in the Amazon (Rocha and Pereira 2003). Migratory movements are different for medium-size dams. According to Rocha and Pereira (2003), the Amazonian migratory flow observed in Tucuruí caused a demographic pressure which resulted in a social and economic imbalance.

There is little information and work on migrations and in particular those of workers on hydroelectric plant sites. Companies communicate very little on the subject and do not let academic researchers do in-depth analyses of their employees. However, we estimate that large dams, such as Tucuruí or Jirau, employ between 7000 and 10,000 people at their peak. Very large works, such as Itaipu or Belo Monte, can have more than 30,000 people on site. Hydroelectric plant workers are difficult to question, but some studies exist on their lives (Lima 2013 and Ribeiro 2013). Very little is known, however, about their origins and their movements (Cavalcante 2005; Alves and Junior 2012). There is a lack of information around this topic, which represents a field of research to explore.

Hydroelectric dam workers have a specific name in Brazil: the *"barrageiros."* They specialize in this type of construction and move from one project to the next. The term *barrageiros*, somewhat pejorative depending on its use, incorporates the notion of displacement and accompaniment of dam sites in Brazil. We will discuss later (Cf. Sect. 2.4.1) the question of identity of the *barrageiros*. They are a dynamic community on the Internet, forming groups of information and support, communicating to keep themselves informed of job offers and upcoming projects. They are present through all the phases of construction.

The jobs that accompany the construction of a dam are diverse and require all manner of skills. Whether they are highly qualified engineers or simple laborers, all workers can apply. Apart from construction, there are jobs in the food department, miscellaneous services, security, and so on. The sudden high demand for workers attracts a large number of people, even within the municipality where the dam is built, often causing migrations at local and regional level. Rural populations leave their place of residence to come and look for work in urban areas. Pille da Silva (Silva 2005) describes this phenomenon for the Itá Dam. In addition, there is an economic migration of workers in rural areas almost automatically from the moment the building of the dams begins, where local labor is not sufficient. Finally, with the progress of construction, the demand for employment can change as the tasks and work available evolve. The type of jobs changes, creating intense and diverse migration flows throughout the construction process.

The recruitment by the company is almost constant, and it does not necessarily need to put in a lot of effort to find workers: The offers are published both locally and nationally on all types of devices. On connected job vacancy platforms, dam sites are often very attractive and offer many possibilities.

In addition to formal and official jobs from the company and its subcontractors, the informal job market also grows around the site. Informal employment, which is still very much present in Brazil, can concern both illegal activities related to drugs and sex, for example, but also those of services and businesses. The financial windfall that arises can cause a loss of control of the economic situation by the institutions and thus an increase in illegal activities. A real lack of control by the authorities is evident. Once a dam is built, local police are more or less assigned to protect the structure and its workers, but they cannot channel the arrival of people who act illegally around dams. The informal sector unfolds in the peripheral area of either the nearby urban logistics center or the dam site. In both cases, there is a search for proximity with construction site workers, who are potential customers of the illegal market.

With the large influx of workers on construction sites, the company must take responsibility for housing them because there is often no necessary infrastructure on the site of the dam and/or in the nearest town. The number of incoming workers is sometimes greater than the local population. The company then builds housing, sometimes even complete neighborhoods and/or small villages in order to temporarily accommodate workers and their families and provide for all their needs: schools, shops, health, safety services, etc. These are the services of a city built from scratch. In addition to housing families, these infrastructures provide a certain stability for the workers, thus avoiding excessive turnover and population movements. The largest dams, like Belo Monte, Sobradinho, or Ilha Solteira, are surrounded by small villages that can become cities. The example of the municipality of Ilha Solteira/SP (São Paulo) is interesting because the transformation of the village built for dam workers into a real organized and planned city did not change the original socio-spatial segregation (Lima 2013). In the case of Sobradinho, workers remained on the site after completion of the dam and helped to create a proper living environment, which officially became a municipality in 1990.

The economic migrations that accompany dam constructions are structured differently depending on the migrants. The most important flow is well organized because it concerns the site workers who are often specialized in civil construction. On the other hand, an uncontrolled flow of population looking for work or seeking to take advantage of the particular context also deeply marks the territory, possibly with a loss of control. The socio-spatial disturbances begin with the massive arrival of people that cannot be sufficiently managed by the company in charge of the works.

Forced migrations are a second major type of movement around dams. They are one of the main negative social effects of the construction because families are forced to change their place of residence. Forced migration, which may be justified or not, has a profound impact for families from an individual point of view but also a collective one for society, because the displacement of entire groups of people profoundly alters the territory and its dynamics.

Forced migrations generate debates and reflections due to their complexity and diversity. They are generally divided into several categories as described by the Refugee Studies Center (Department of International Development, University of Oxford). The different types of forced migration are as follows:

- Conflict-induced displacement: People who are forced to flee their homes because the state authorities are unable or unwilling to protect them when they have to face an armed conflict generalized violence or persecution on the grounds of religion, race, political opinion, etc.
- Development-induced displacement: People who are compelled to change their place of residence as a result of policies and so-called development projects such as large-scale infrastructure projects (roads, airports, dams, mines, etc.).
- Disaster-induced displacement: This category includes people displaced as a result of a natural disaster (hurricane, earthquake, etc.), environmental change (desertification, climate change, etc.), or human-made disasters (industrial accident for example).

These types of forced migrations produce migratory flows composed of various types of migrants, which can be categorized according to different criteria. The Refugee Studies Center lists seven types of migrants: refugees, asylum seekers, internally displaced persons, development displacees, environmental and disaster displacees, smuggled people, and trafficked people.

Our study concerns the "development displacees," who are sometimes also referred to as "oustees," "involuntarily displaced," or "involuntarily resettled." This migratory flow is often important in terms of the number of displaced persons but receives less attention and assistance than others. It generally affects poor people, urban as well as rural areas, and indigenous and ethnic minorities. The Refugee Studies Center estimates that during the 1990s, between 90 and 100 million people around the world were displaced because of the construction of development works.

The terminology used is questionable because the term "forced" depends on the origin of the migration (Turton 2003). We can talk about compulsory or coerced migration in contrast to other types of migrations that are voluntary or at least decided. Thus, in the case of migration caused by development projects, such as dams, or natural disasters, we will talk more about compulsory migration. Van Hear (1998) explains migrations according to a matrix made up of two axes. The first axis represents voluntary (more choices and options) and involuntary migrations (fewer choices and options). The second is built on five movements: inward, outward, return, forward, and sedentary. For involuntary migrants, he cites refugees and development-induced migrants, including those affected by dams. The lack of options and choice defines this migration flow. The United Nations Development Programme (UNDP 2009) states that "it is rare, not to say exceptional, that choices are entirely free" and that "displacements often have harmful consequences when they occur under conditions of limited choice." These migrants do not decide when, how, and where to migrate. In addition, they are in the hands of the company who makes the basic choices for the migration. These two components characterize forced migration: being compelled to leave one's home and not having options and choices to model the migration as one pleases.

For victims of dams, forced migration means expropriation. Legally recognized by Brazilian law, it allows to move families on the basis of a declaration of public utility. Specifically for the electricity sector, the declaration is issued by the sector's regulatory agency, ANEEL. Since its creation in 1996 and in the context of privatization of the sector, ANEEL is responsible for this competence, which is not without debates (Ferreira 2009). The declaration is based on Article 29, Paragraph VIII of Law 8.987 of 1995, combined with Article 10 of Law 9.074 of 1995, with the wording given by Law 9.648 of 1998 and regulated by Resolution 279 of September 11, 2007. According to these terms, the expropriation process takes place in two phases:

- A declaratory phase, for which a delegated entity (construction company) expresses an interest in purchasing assets (lands and properties) in order to carry out an activity of public interest, under the law. The company already knows where the impacts of dam construction (dam and reservoir) are located.
- An executory phase during which the measures necessary for the transfer of assets are taken (Ferreira 2009). People are displaced, and the company legally acquires land and properties.

The request for a declaration must be made by the company responsible for the work that will produce electricity (in our case the hydroelectric plant). To obtain the declaration, the company must meet some requirements (disclosure of information to the affected population, publication of criteria for compensation, presentation of negotiations with owners for example). But, with the 2007 resolution, the request for declaration was simplified (in comparison with the previous resolution of 2003) and the requirements became mandatory a posteriori. The requirements are therefore not really necessary to obtain the declaration (Ferreira 2009). This simplification is proof of the influence of the lobbying by the electricity sector within the Brazilian political power. This change indeed greatly simplifies the process and leaves the door open to all types of treatment toward dam-affected populations. Finally, the same author also asserts that ANEEL goes beyond its institutional competence and there is a risk that the institution will promote typical agrarian reforms, even if this is not evident or proven.

Measures are included in the law to compensate for suffered losses and for the expropriated to recompose their assets (Carvalho 2006). The types of compensation offered are diverse and again specific to the electricity sector.

It should be noted that the forced migrants are the inhabitants of the lands that are affected by dams, whether they are owners or not. Living in an area that is subject to the declaration of public utility forces the families to abandon their residence and to find another. Before the Itaipu Dam, it was generally only the landowners who would receive compensation. With the emergence of social movements to defend dam-affected populations, families renting or simply occupying agricultural land also started to receive compensation from the companies. In addition, agricultural land affected by the dam but not occupied by people also falls within the compensation criteria for its owners.

Forced migration is virtually systematic for lands flooded by the dam because, in addition to rising waters, there is a strip of land (Permanent Protected Area—PPA) around the reservoir that must be owned by the dam company and where families should not be able to settle. However, there are still cases of families living along reservoirs, thus showing the companies' weak control of the territory. In addition, there are often negotiations on the boundaries of the reservoir and the flooded lands. In some cases, wealthy owners have indeed managed to keep their residence after negotiations with the company, despite the proximity of the reservoir. Forced migration is therefore also subject to power struggles and economic inequalities.

Forced migration mainly concerns populations in rural areas but sometimes also in urban areas. The latter are, however, less impacted because dams seldom affect urban spaces, due to the infrastructure and much larger population density.

Forced migration is a systematic outcome of hydroelectric plants. Local people pay the price of energy development through a compulsory change of residence and place of life. Negative and/or positive effects are more measurable in the consequences of migration than in migration per se. The compelled change of residence sometimes leads to other migrations.

Various migratory flows of different natures can follow one another in time. For instance, other migration flows linked to the construction of hydroelectric plants may follow forced and economic migrations. These are mainly return migrations. The International Organization for Migration (IOM) defines a return migration as a return to the place of origin of the first migration. It is therefore the second phase of a migratory process that has already begun. The time frame considered by the IOM is about 1 year, but this criterion remains rather arbitrary. Return migrations related to dam construction are diverse as they can concern all people who have already migrated because of the project. The time criterion is not taken into account because the return movements tend to be faster. A true return migration mainly concerns workers. Otherwise, there are also attempted return migrations by *atingidos*, but these cannot succeed as their original land and home no longer exist.

In our study, we found that return migration was most common for employees of the dam company. Once completed, a dam cannot accommodate all the employees who worked on its construction. Only a small number remains to operate the plant on a daily basis and to control the surrounding area. The workers then return to their places of habitual residence or go to another dam job. After spending several months or years on site, they can return to live in their region of origin. Reintegration, a fundamental element of return migration, needs to be done in a positive way. As the original migration was economic, we indeed assume that return migration is accompanied by a certain money saving. This return migration does not, however, always occur because some employees (*barrageiros*) move from one dam construction site to another. They then have to wait several projects before returning 1 day to their region of origin. Finally, other employees decide to stay on the construction site of the dam and settle there. This is the smallest contingent, but this phenomenon exists for almost all dams as observed for small dams (Dona Francisca) as well as for the most imposing (Sobradinho).

The second type of return migration concerns dam-affected populations forced to migrate initially. Although it does not exactly fit the criteria, it fits in spirit, with the idea of returning to the starting point of the first migration. This return migration can be motivated by various elements: a failed or difficult adjustment, a search for lost links, the economic market, etc. It can happen quickly after the first migration in the case of families who received compensation they did not really want. We often find abandoned or sold homes in family resettlement areas only a few weeks after the inbound migration. Even though it is illegal due to an agreement of a minimum time of permanence on the new place of life, the sale of new houses built for dam-affected populations exists and modifies the territory planned by the company. The return migration of dam-affected families can also be an escape from an uncomfortable situation and the search for a known and more adapted situation where they can find their bearings (social, territorial, economic, and spatial).

One last return migration that we want to address concerns the more volatile populations attracted by dams to take advantage of the opportunities that come with them. Dabbling in illegal activities or not really settling in the area, these populations migrate to new opportunities without any planned agenda.

The types of migrations that accompany the construction of hydroelectric plants are diverse and evolve over time, following the pace of construction. Economic migration takes place before and during construction; forced migration of the local population happens shortly before completion of the dam; return migration takes place at the end of construction or after. The complexity and diversity of the migratory flows show the variety of population movements in the region of construction.

2.1.3 Profile of Forced Migration

Forced migrations flows raise many questions: Who are the migrant populations; what are the migration flows; what are their destinations, etc. Forced migrations concern dam-affected populations, referred to as the "*atingidos*" in Portuguese. We even speak of direct *atingidos*, in opposition to indirect ones who suffer from consequences other than expropriation. It is easier to analyze direct *atingidos* because we have more information about them than for the other migration flows described earlier. Some studies exist on other types of migrations, but they address the economic and social consequences and not the flows, i.e., the number of people, the stages, the destinations, etc. We will try to identify patterns or models of migration flows.

The study of human migrations often involves an analysis of numerical information, numerical data on the number of people, places of departure and arrival, etc. Information on the victims of dams is scarce and difficult to access for a variety of reasons.

First, primary data comes exclusively from the private sector (companies responsible for the dams) or previously, during dictatorship, from the public sector. Official information on migration sometimes circulates but only global data or only on specific types of compensation. In comparison with the technical information of the dams, the amount of social data available is insignificant. Raw data exists but never reaches the researchers. We must therefore look to other sources, including social movements. Like the companies, these organizations are less likely to communicate information because they are influenced by the political game and by the lack of resources available to them. Public institutions do not give access to data either, despite having it, notably through environmental licenses that force companies to provide encrypted information. Finally, academics can produce data, but only through surveys (Leturcq 2012).

Another difficulty lies in the variability of the terminology used depending on the object studied, such as the terms "families" and "people." Companies talk about the victims of dams as families, but social movements and other actors talk about people. The difference comes from the fact that the number will be much larger and therefore more striking if we refer to victims as people. To better understand, the number of people per family per household has long been considered to be five, despite having decreased to 4.5 in 1980 and continuing to decline since. In rural areas, it was still 4.7 in 1991 and it went down to 3.6 in 2010, while for the country it was 3.3 (IBGE 2010). It is therefore necessary to pay attention to this in readings and analyses. We will mostly use the family reference because it is the most commonly used and the one for which we found the most information.

The profile of the victims of hydroelectric dams is difficult to establish. As dams are present throughout Brazil, it is difficult to draw a unique profile of the *atingidos* because the affected populations vary depending on their location. We will present a general profile, mainly of the rural populations, supplemented by information obtained from our direct research among families. In 2007, we interviewed families affected by the Machadinho Dam who had resettled in collective rural *reassentamentos* and families who were soon to be victims of the Foz do Chapecó Dam. Our study was completed with families affected by the Belo Monte Dam who had resettled in the new district, the collective urban *reassentamento* of Jatobá in Altamira/PA.

Dam-affected populations are largely rural because dams are established in valleys and affect agricultural areas, but some urban areas and even cities can also be affected by the construction of dams, such as Altamira and Porto Velho, respectively, with the Belo Monte and Santo Antonio dams. Rural areas are diverse in their geo-climatic, morphological, landscape, demographic, and other characteristics. Affected rural populations mainly consist of small farmers (80%, for example, for the victims of Sobradinho (Sigaud 1988) producing for the local market or for their own consumption. The description of the Brazilian rural population based on census data gives a better idea of the dam-affected population.

Rural populations tend to be older than the rest of the country because Brazil has been suffering from a significant rural exodus since the 1960s as young people move to urban areas looking for their first job (Maia and Buainain 2015). The aging process of rural areas is related to their dwindling population. According to IBGE censuses, there were 1.6 million people aged 65 and over living in rural areas in 1991 and that number increased to 2.2 million in 2010. The population aging index rose from 9.7% in 1991 to 21.4% in 2010, especially in the more traditional rural areas of the south, southeast, and northeast (Maia and Buainain 2015). The average age of the rural population in Brazil was 30.6 years old in 2010, up from 1991. The state of Rio Grande do Sul has the highest average age for rural populations at 37.2 years old, and the state of Amazonas is at the other end of the scale at only 24 years old. *Atingidos* interviewed from the south average 28 years of age at Machadinho and 30 years of age at Foz do Chapecó. The average age of the 1127 individuals interviewed in the neighborhood of Jatobá in Altamira was 28.4.

The aging of the rural area also occurs with changes in the family profile, with a reduction of the fertility rate and the number of members per family. In 1991, rural families had 4.7 people on average, while in 2010 there were only 3.6. The fragmentation of families can be explained by the rural exodus but also by the decline in fertility. The fertility rate indeed drastically decreased overall in Brazil, going from 6.3 children per woman in 1960 to 2.4 in 2000 and 1.7 in 2015 (IBGE 2010). This drop can be explained by the usual reasons, mainly more accessible education, an improvement in the types of contraceptives available and a general improvement in public health.

At the same time, there is also a masculinization of the Brazilian rural areas for all age groups, which is also a consequence of the rural exodus. In 2010, 52% of Brazil's rural population were men. We can confirm this phenomenon from our interviews in the south, with 53.6% of future *atingidos* of Foz do Chapecó being male and 52% for the *atingidos* of Machadinho. The population (50.4%) in the district of Jatobá is male. The male presence is evident in the studies on dams because, in addition to being a majority in rural areas, men are also clearly the majority in civil construction. According to the Annual Relation of Social Information (IBGE 2016), in 2015, men represented almost 90% of the sector's labor force.

Some other elements characterize the Brazilian rural population, such as religion. According to the 2010 census, 77.8% of the rural population were recorded as belonging to the Roman Apostolic Catholic Church. This percentage is significantly higher than the national level (64.6%) but down four percentage points compared to 2000. The second most represented religion in rural areas is the Pentecostal Evangelical Church (10%) which gained almost three percentage points in 10 years.

The importance of the Catholic religion is illustrated in the structure of rural areas. The church indeed represents a central component of community life, especially in southern Brazil. For the *atingidos* of Foz do Chapecó, out of 41 responses, a little less than half (19) declared themselves Catholic, attending the local church. Non-respondents to the survey identify themselves as Catholics but non-practicing. In Altamira, 63.7% of dam-affected families say they go to church and 30% attend the Catholic Church. There is a great diversity of attendance because more than twenty churches are cited. Some families complain about the lack of church in their new neighborhood, while others indicate going to any church.

In terms of skin color, according to the self-definition of the population (IBGE 2010), 54% of the rural population declared themselves mixed-raced, compared to 43% at the national level. The second determination is white for 36% of the inhabitants of rural areas, against 48% overall in Brazil. Finally, Indians are more numerous in rural areas, representing 1.6% of the population, against 0.4% at the national level. Based on observations, *atingidos* consist of a higher proportion of mixed-raced people and "caboclos" (individuals born of Indian and European descendants). The location of the dams, in valleys and remote areas, is an important factor here because the poorest rural people, including the indigenous people and the *ribeirinhos* (riverside dweller), especially in the Amazon, occupy these areas which are difficult to exploit and whose land value is lower than in the plains and plateaus.

In terms of literacy, rural Brazil is clearly behind the national level. More than 21% of the rural population aged 10 or older was illiterate in 2010, compared to just 9% nationally. This gap can first be explained by the difficulty of access to the education system. The issue of access to the school system for the *atingidos* is linked to the location of the dam. Among the heads of dam-affected families in Altamira (409 people), 11% never went to school and 46.5% did not complete primary education.

This enables us to draw a broad portrait of *atingidos* in Brazil. The profile is therefore a man, about 30 years old, Catholic, mixed-raced, with a family of less than four members, and working in the agricultural sector in a rural area. His financial status is limited, and his level of education is lower than the average of the Brazilian population.

Time is an important factor in forced migration due to the construction of dams because not all affected families migrate at the same time and at the same pace. Several stages are noticeable in this migratory phenomenon.

The first forced migration occurs outside of the legislative framework and is neither forced nor supervised by the dam company, but it is nonetheless a direct consequence of the dam construction. It is an anticipated migration because it precedes the work. Some families or institutions do not want to wait to see the dam come out of the ground to leave the region and rebuild their life or projects. The simple announcement of it can trigger a migration which has the advantage of being thought and planned by the family. The anticipatory migration is compelled by the perspective of the dam. It is nevertheless governed by the families' own rules and not by those of the company. This type of migration thus has characteristics

the two zones: rural–rural, rural–urban, urban–urban, and urban–rural. This order follows the importance of the migratory flows observed.

As far as distance is concerned, our study only focuses on the migrants we can follow up on, that is those who chose compensation in a collective *reassentamento* or those who received an individual compensation and whose information was shared by the dam company. Based on agreements between the companies and the *atingidos*, the relocation distance is sometimes determined before the migration. Pre-agreements sometimes exist to establish a maximum perimeter. In the south, an agreement specified a maximum distance of 200 km for the collective rural *reassentamento*. In the north, Norte Energia established a maximum radius of 2 kilometers to resettle the urban populations affected by the Belo Monte Dam, in Altamira. However, these guidelines are not necessarily respected and, in both cases, we found examples of families resettled beyond the limits. For older dams, the notion of distance did not matter as dam-affected families were not generally acknowledged. For example, more than 1000 families affected by the Sobradinho Dam had to move to a new home about 700 km away (Sigaud 1988).

The victims of dams, who have no other choice but to migrate, wish for the vast majority to stay as close as possible to their former place of life. As explained by Sigaud (1988) and proven in field research among the *atingidos* of Machadinho and Foz do Chapecó (Leturcq 2010), proximity is fundamental and it is the first criterion for families. In 2007, we analyzed the locations considered (18 municipalities) in order to place victim families of the Machadinho Dam in *reassentamentos*. The average distance to the dam was 109 km (Leturcq 2010), and the locations were in the three southern states of Brazil.

We can differentiate the migratory flows according to the compensation received by the families. Those who receive money in compensation have the choice to migrate wherever they want, and therefore we have no information for them. For the *reassentamentos*, collective or individual, it is sometimes possible to get information on where they migrated to.

Among the families affected by the Machadinho Dam (2002), 183 opted for an allotment in a collective rural *reassentamento* (of which, more than 45 families were in an unofficial *reassentamento*) and 764 for an individual *reassentamento*. There are five collective *reassentamentos* in the two southern states (Santa Catarina and Rio Grande do Sul). These are generally located within 70 km of the dam and the furthest, in Curitibanos/SC, is about 120 km away. We indirectly obtained information on migration flows for 397 families in individual *reassentamentos*. The first characteristic is the proximity of the departure zones: 95% of the families came from three municipalities near the dam (Leturcq 2010). Next, more than 28% of the families moved to another location in their municipality of residence. Thus, 114 families managed to stay in the same locality, only a few kms from their old place of life. Finally, despite the dispersal of the migration destinations (22 municipalities), the great majority of migrants only moved a short distance, about 35 km from the Machadinho Dam, mainly in the municipalities neighboring their original municipality.

For the older Sobradinho Dam, information comes from the comprehensive work of Sigaud (1988), inspired by that of G. Duqué. The number of victims varies between 60,000 according to the company (CHESF) and 72,000 according to the union of rural workers, or 11,853 families according to a CHESF document from 1980. Sobradinho is the dam that caused the most forced migrations. The urban victims of the reservoir (3234 families according to CHESF) had to move to four new urban centers built in the region and did not opt for rural areas. For rural populations, the scale of migration and the lack of political will to manage the situation caused chaotic and sometimes surprising situations for families. The migration options varied with time (Sigaud 1988; Duqué 1984). The first option offered was the "do-it-yourself" option where the company only offered a bus ticket and compensation for the family's assets. About 2300 families chose this option and often decided to only move a few kilometers away, still in the area soon to be flooded. Observing this move, the company decided to offer another option for families, the "lakeside" option, and established 25 rural communities near the new reservoir to house 2600 families. As a final option, the company resettled about 1800 families in the nearby caatinga. At the same time, almost 1000 families decided to leave the local rural area to head for new towns, also near the reservoir. For the rest of the rural population (about 1000 families), the company offered the option of resettlement through an INCRA settlement project, 700 km from the Sobradinho Dam, in Serra do Ramalho. Families who chose this option found it hard to adapt because they would have preferred to stay near the reservoir and their old place of residence (Estrela 2007). The dam dates from a bygone era, but even then there was already a desire from the affected people to not move away from their former place of life and to preserve their way of life and their social and territorial relations.

The Belo Monte Dam affected both rural and urban populations. The urban victims of the Belo Monte Dam represent the largest contingent of urban victim families of a dam. According to the latest Norte Energia reports (May 2017), around 9,000 families affected by the dam had to migrate. Of this total, 4000 chose cash compensation and 3,600 resettled in six collective urban *reassentamentos* (RUC) on the outskirts of the city of Altamira (Cf. Fig. 2.1). The Pedral RUC, more than 6 kms south of the city center, is also the smallest (150 houses), the last built and intended for urban Indian families and *ribeirinhos* with an easy access to the river. All RUCs have houses with a small garden and somewhat developed collective infrastructure: school, bus services, community center, dispensary, electricity, sewers, garbage collection, sports fields, etc. They accommodate between 1150 (Jatobá) and 450 (Casa Nova) houses for a total of 3850 houses built. We have observed that these houses are not all occupied, quite the contrary, and some are already up for sale even though this should not be legally possible. In this case, the migrations are twice forced because families have to move first from the departure zone and later from the arrival zone. Families who move to an RUC do not have a say in the location or choice of house. The feeling of being "taken for a ride" is clear from interviews with *atingidos*.

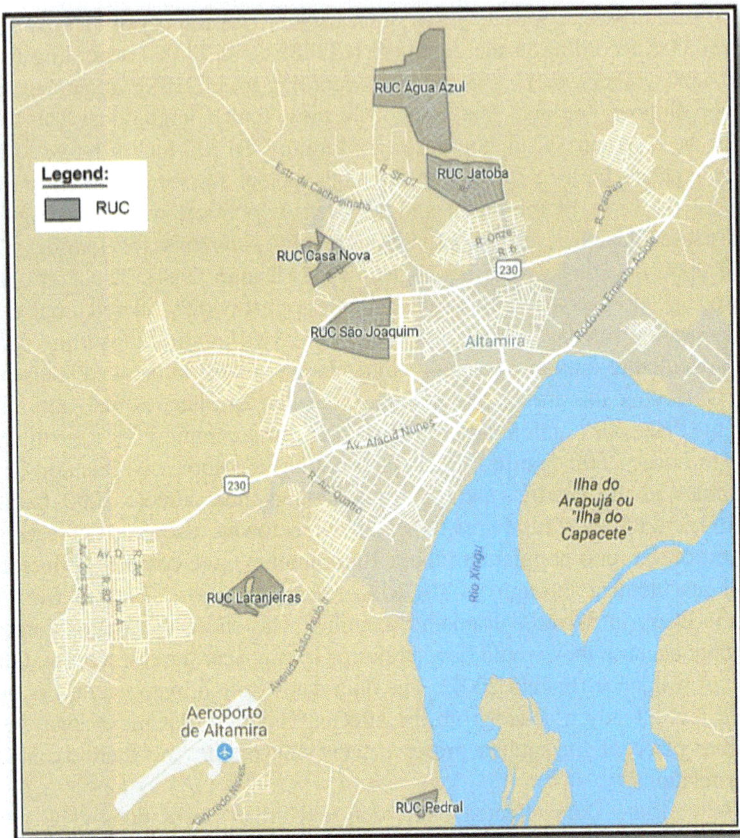

Fig. 2.1 Localization of collective urban *reassentamentos* in Altamira. Modified by G. Leturcq, 2017

The three previous examples, and many others, make it possible to construct a diagram of forced migrations due to dam constructions in Brazil. The flows of families coalesce between the two zones of departure, rural and urban, and in three zones of arrivals, rural, urban, and unknown. We find migrations between the two main zones, in both directions, and within the same zone. All possible migratory scenarios exist for *atingidos*, and there is still an unknown migratory space which has not been registered nor documented. In the following diagram, we illustrate the various flows and the main migratory areas. Across the region affected by the dam, the urban areas are punctual spaces: small villages or city districts. In rural areas, migratory flows are more diffuse and the affected areas more zonal. There is a wide diversity of migratory flows but spatial logic prevails because the most sought-after area is always the one closest to the river, i.e., the starting point for migrations. The first zone is prohibited by a federal environmental protection law (APP). The second zone is very difficult to access due to a sharp increase in land value and a lack of land available. Finally, the third zone (limit near area) is the most sought after because it is located near the original place of residence. This concerns the municipality of departure or the neighboring municipalities. Then, there is a larger area of migration where families settle more pragmatically, often for questions of price and availability. Finally, there is an unknown area where dam-affected families, rural and urban, can choose their new residence, near or far, thanks to their monetary compensation.

We must therefore acknowledge the existence of a whole migratory system that accompanies the construction of a hydroelectric plant. The complexity of this system lies in the fact that it is made up of various populations, that the migrations occur for different reasons, that the spaces concerned are either urban or rural, and that the flows go in all conceivable directions. A specific migratory model of *atingidos* comes out of the analyses, in particular according to the place and the conditions of departure. The flow is then influenced by the choices and options available to the families. The migratory system deserves special attention for future research, and it is important to note that the continued construction of dams will continue to result in the movement of people. Finally, we must acknowledge that this migration system is part of the long history of migrations in Brazil and participates in the continuous movement of the population in the territories (Fig. 2.2).

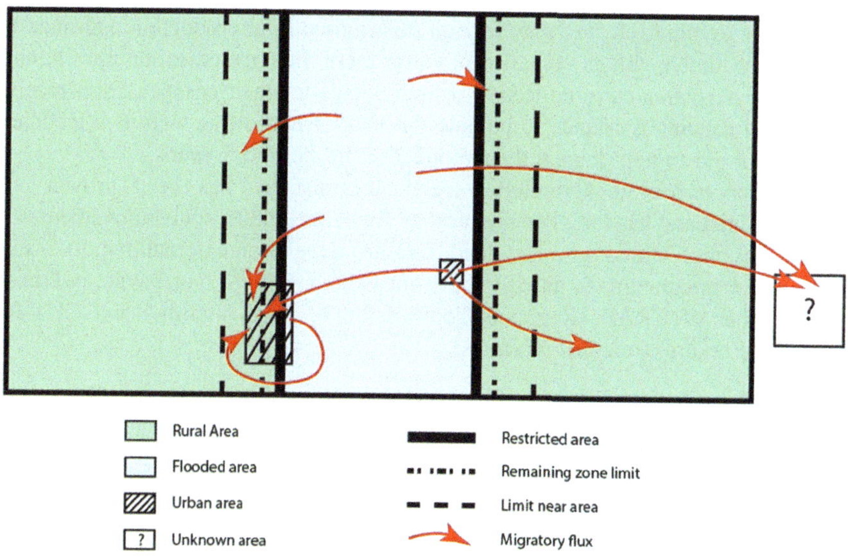

Fig. 2.2 Diagram of migrations near a dam. Created by G. Leturcq, 2017

2.2 Modification of Living Space

2.2.1 Economical Sphere

The construction of a dam has serious consequences for families living in the area, such as forced migrations. Families must change their place of residence, as well as their productive activity or profession, and adapt to a new environment. With the dam, agricultural lands are flooded, which alters the economic profile of the region. With changes in the labor force and the productive environment, part of the economy is shattered at the margins and must be restructured around migrant families.

Forced migrations oblige families to rethink their lives. They face a choice: to maintain the same lifestyle or to change it radically. Even if they choose to carry on in the same way, by moving to a similar place, they will still face various structural changes: a new house, a new living environment, a new social structure, etc. The changes are numerous and affect the families' productive activity. Most victims of dams are small farmers dependent on commercial or subsistence farming within a limited radius. These families, who depend on the land but also on the nearby river, have a strong link with the area. The land–water relationship exists for all dam-affected families and evolves or disappears as a result of the migration as it is rare for families to be able to resettle near a watercourse.

Migrating means having to adjust to a new land as well as different geomorphological and climatic conditions, and therefore it is also necessary to adapt agricultural practices. Small farmers need to change their methods, their productions, their rhythms, etc. As they rely on their agricultural production and must be quickly productive, this is very disruptive for them. In some cases, families having chosen a *reassentamento* (collective or individual) for their compensation receive an initial financial assistance to palliate the lack of harvest as well as a technical agricultural aid to help launch their production in the first 5 years.

The importance of agricultural technical assistance has been proven and acknowledged and has long been requested. Since the 1980s, social movements are aware that, once settled on a new land, families depend on external help to ensure their proper integration. In an editorial from February 1988, Ivar Pavan, politician and member of CRAB (Regional Commission of Dam Victims) and CUT-RS (Unified Workers' Central), indicates:

> We must ensure our permanence on the land after the construction of a dam. We must therefore implement techniques adapted to the type of agricultural property we own and also organize political fights to guarantee compensatory prices for agricultural products, lower interest rates and good health conditions and education. We will then be able to produce more and live better (CRAB 1988).

Dam victims sometimes move to a productive land which is radically different to where they came from. If they lived on a riverbank, in a hilly area and therefore on sloping land, they sometimes choose to settle in a plain where crops are different. The proximity of the river modifies the relationship with the water resource, and therefore its absence also plays a role. We have seen families adapting to the environment of a southern *reassentamento* by getting together to buy a tractor. Solidarity and mutual aid enable families to make investments. This is a new practice for those who did not have access to these techniques and funding previously. Other families opt for irrigated crops. They invest in the material and learn how to use it to produce more. The transition from a self-sustaining agriculture to a commercial mechanized agriculture is common but far from systematic. By doing so, there is a real risk, more or less induced by migration, of not being able to turn back.

The integration of families into the new economic framework also implies their entry into the local market. Whether involved in the productive or commercial process, families must learn to control their new environment. A good agricultural production calls for good relationships with salespeople for seeds, fertilizers, herbicides, etc. A new set of products and techniques must be mastered, and this requires establishing new relationships with local stakeholders. Once their production is established, the producers must enter the local commercial circuit. Buyers and unions have key roles in selling products. The dam-affected families must therefore find their place in the territory of the local economic stakeholders.

A collective *reassentamento* represents a positive structure for the adaptation to the new economic framework because the mutual aid and the division of the tasks, costs, and investments offer possibilities that families who choose to resettle individually do not have. The new community created within the collective *reassentamento* plays an important role in the new economic life of the families. Whether rural or urban, we noticed a real economic dynamism in the *reassentamentos* we visited throughout Brazil.

The example of the collective urban *reassentamento* of Jatobá, in Altamira, speaks for itself. In just under 2 years, we observed a proliferation of economic activities in an area that had been created from scratch. Among the thousand houses offered to the families affected by the Belo Monte Dam in October 2015, we counted 178 service and trade establishments and 6 street vendors. All business sizes and all activities are represented. For example, a family sells ice cream from home, while their neighbor has opened a bakery and a little further away an at-home car wash station can be found. The diversity of activities is striking because it is almost spontaneous. Norte Energia, who is responsible for the dam, has helped out a few families, but the vast majority have built up their business from what was left of their compensation combined with their desire to start working quickly. These spontaneous businesses also help the families to participate in the creation of a community in this new peripheral district. From a spatial point of view, the services and shops are found in all the streets, with a slight concentration along the central artery that divides the district in two. Finally, we must point out the presence of churches, a dispensary, a school (under construction in 2015 but completed in 2017), a sports field, a community center (2017), and a city bus stop. The new economic life of the district works in synergy with the dynamism of the population and the infrastructures created by Norte Energia.

The economic growth of dam-affected families is difficult to assess because it is extremely varied and has not been studied. A specific study of the fishermen nevertheless allows us to obtain a more detailed overview of their economic development. Fishermen form a well-defined category facing a rather complex situation during the construction of a dam because of the difficulty to control the profound alteration of the water resource, as demonstrated by the extensive research carried out on the reservoir of the Tucuruí Dam (Mérona et al. 2010; Flexa et al. 2016; Almeida 2016). The repercussions of dams on the fish population are proven, with a general reduction in the number of species and a decrease in production (Mérona et al. 2010). After years of protests and hardships, the first supportive public policies have come out in an attempt to help fishing families, both upstream and downstream of the dam, which is unprecedented.

The lack of studies on the actual economic effects on populations prevents us from drawing conclusions with absolute certainty, but our field observations have helped to establish that positive and negative situations emerge from the construction of a dam. Some families adapt to their new living environment and thrive, economically and in general, while others find themselves in precarious situations, not to say poverty, sometimes worse than before the arrival of the dam. We observed, in the same neighborhood, situations of complete failure besides great successes (e.g., in the *reassentamentos* of Campos Novos/SC or Altamira/PA). Some families compare the arrival of the dam in their lives to winning the lottery. Others take advantage of the opportunity to change their life, sometimes returning to live in rural areas, for example in Altamira. Families migrating from an urban to a rural area often have the chance to become land owners, sometimes for the first time. The notion of opportunity often comes up in interviews with families. The picture of the economic situation of families after forced migration must therefore incorporate both the negative and the positive.

It would be interesting to complement the research with a regional project to assess the economic changes encountered by dam-affected families and then to try to understand the key factors in the success or failure of adaptation to the new environment. We can make some explanatory assumptions, among which: the burden of the migratory history, the migration process, the participation of the family in the decision-making process, the financial situation at the migratory destination, and the sense of justice.

2.2.2 Community Space

Dam constructions profoundly disrupt rural and urban areas. They affect community structures as well as the spaces organized by the populations and their social networks. Migrations profoundly change the relationships between families and thus the structures of existing communities. In return, new communities emerge, built around a common idea to redo one's life after the forced migrations. Rural areas throughout Brazil are home to small communities living close to rivers. In the

Amazon, as in the semiarid northeastern region or the temperate south, valleys are occupied by people who take advantage of the water resource and the location. In the north, on the banks of the Xingu River, *ribeirinhos* are settled in small clusters. In each area, there are only a few families living together, on some islands of the river. In the south and northeast of the country, a few isolated houses can be found but families generally tend to gather around a focal point (religious, economic, cultural, school, etc.). After looking at the community and the importance of social links, we will describe three types of communities that must evolve as a result of a dam construction: abandoned communities, new urban communities, and rural communities.

The economic activity of a family does not suffice to define its living environment; its social connections also need to be taken into account. In a rural area affected by a dam, communities play a very significant role because they can provide support for families. Their weakening is therefore likely to have negative consequences. We use the term community based on Henri Mendras' interpretation: "Peasant societies (…) are organized into relatively small and autonomous communities settled on a territory that they exploit" (Mendras 1995). Communities consist of families linked by neighborly relations and/or family ties. According to Angeon (2008), the community's territory is developed with "the natural inclination of members to create relationships and organize themselves to collectively engage in actions." Good relationships and organization are essential for the construction of a collective structure and to allow a satisfactory evolution of the territory.

The rural areas around the future reservoirs are made up of small communities where many relationships are established. These communities are built around social bonds (families, friends, neighbors, etc.), cultural associations (leisure, worship, etc.), and services (health, education, transportation, etc.).

Families nurture close relationships between people. In remote rural areas of Brazil, we met families with complex structures including collateral relatives and mixed generations. In all parts of the country, we interviewed families living near siblings, parents, cousins, and so on. Family structures evolve within the same community or in close communities, often in the same municipality. The spatial morphology of communities is governed by the watercourse which also helps in bringing families together (Leturcq 2010). Houses are close to each other, and individuals sometimes gather for activities related to the river (recreation or fishing for example). Communities can develop along the watercourse or gather a few hundred meters from it and organize themselves around a small center (church or community center). Both types of spatial organizations can be seen in the municipality of Rio dos Índios/RS, near the dam of Foz do Chapecó.

In questioning the soon-to-be *atingidos* around Foz do Chapecó, 46 of the 65 heads of families claimed to have family living nearby, in the municipality. In fact, 20 of them had at least 20 members of their families in the municipality. Family visits usually occur five to ten times a month, but ten families mentioned visiting their relatives more than ten times a month. The relationships are therefore very well established. In Altamira, nearly 60% of the 269 homes of the victims of the Belo Monte Dam visited had family members who had to migrate because of the dam. Both examples show the proximity of families in both rural and urban communities as the latter live in precarious neighborhoods where mutual help is important.

Family is not the only essential link in a community's social relations, neighborhood, and friendship count too. Good neighborhood relations in rural communities are common. For the future *atingidos* of Foz do Chapecó, 46 out of 65 describe their relationships with their neighbors as "good" and only one person qualified it as "bad." Their neighborly relations involve mutual aid, carpooling, agricultural exchanges, and the majority believe they can count on their neighbors in case of health problems.

Shared moments also forge communities. Various group activities help bring people closer such as agricultural meetings, pétanque games, craft activities for women, and especially football games. The main sport practiced in Brazil, football, gives families an opportunity to get together. In a similar way, religious festivals (Saint-Jean's Day for example) and town fairs contribute to the good relations between families. Finally, some illegal activities are also conducive to meeting, such as cockfighting in the south.

The church plays a central role in rural communities and especially with the arrival of dams because historically they have had strong political roles alongside the families (see Itaipu, for example, and the role of the CPT in Brazil). While the Catholic Church seems to be losing some influence in everyday life, it remains a major local player and helps resolve conflicts and social issues.

Finally, the arrival of a dam in a region helps bring together families within the community. Many meetings (on preparation, information, negotiations, mobilization, etc.) are organized before and during the construction of the dam. Although opinions differ on the importance and benefits of dams, these meetings allow families to come together and unite. They encourage the creation of popular and social movements. The reasons for meetings are therefore numerous. They provide an opportunity for rural families to build or strengthen relationships. The fight against dams can therefore be used to unite and strengthen communities.

Rural communities affected by dams experience the loss of members and sometimes entire families. The bigger the community, the more it is affected. When nearly half of the members migrate, those who remain have to adapt and the whole community structure changes in response to departures.

In terms of neighborhood relations, the departures obviously oblige the remaining families to help each other even more. Departures of families lead to further losses, notably of businesses and services, as well as support links. For example, a football team in south Brazil had to look for other players to be able to continue participating in the local championship. Services and businesses are also strongly affected as the municipality sometimes needs to reduce some services for lack of staff (schools for example) and businesses may close for lack of customers.

There is a real psychological shock. A heavy and sad atmosphere emanates from the rural communities affected by the dams. While some disappear completely, others somehow manage to survive. The fragile situation of the community can also prompt other families to migrate to find a better place to live. This can become a vicious cycle of departing families causing more families to depart.

Compared with the pre-migration situation, the paradox is striking, as affected communities are often very active before the dam construction and migration. The usual presence of *acampamentos*, which are occupations of spaces for collective campaign against the dam construction and for claiming fair compensations, reinforces the notion of community but emphasizes the contrast in dynamism (Cf. Fig. 2.3). The camps are organized by social movements such as the MAB and bring together families living in the area who take turns to constantly occupy the site. The location of the camp is often strategic, as for the camp of Foz do Chapecó installed just in front of the site of the future dam. For long periods of time, sometimes more than a year, the camps bring families together and are places for exchanges, meetings, negotiations, etc., that allow the creation of a group and strengthen the unity of the local community. But once the construction of the dam begins and the camp is dismantled, the local community begins its gradual decline.

The community structure then breaks down little by little. For some communities, there is no other alternative than a slow death. The departure of families can often spell the end of a community as observed on the Uruguay River where ghost towns have appeared, due to abandoned communal rural areas, empty houses, abandoned buildings, and fallow land.

Fig. 2.3 Mab acampamento on the Uruguay River. Alpestre/RS/Brazil. Guillaume Leturcq, August 2006

The notion of mutual aid between neighbors is fundamental for these families as they build together a new community space and they have every interest in making it work and prosper. Some families even choose this way of life to be able to count on a trusted neighborhood and a close community to help the process of adaptation. The dynamism of dam-affected families and their social relations is particularly strong because of their life history marked by forced migration.

Rural communities lose a lot of their dynamism and some even disappear, while on the contrary others are born in areas which previously did not have a collective life. The creation of communities plays an important role in the modification of the living space of the *atingidos*. New social interrelationships emerge, built over time in the new neighborhood. At the same time, other familial relationships may weaken and disintegrate, mainly because of the increased distances separating members of the same family after forced migrations.

2.2.3 *Infrastructures*

We have been talking a lot about urban and rural collective *reassentamentos* as they are the only areas where we can meet a group of families affected by the same dam, which facilitates our work to interview migrants. *Reassentamentos* offer a collective answer to a problem of mass migration and better conditions for families, but they are an option often neglected by the companies because they are more expensive and also require more effort and a lot more time. In this section, we will look at the collective rural and urban *reassentamentos*, their infrastructure, and their spatial organization.

The rural *reassentamento* is an option sometimes offered to dam-affected families, but it is far from being the dam companies' favorite and it is not a systematic solution for dam victims. Rural *reassentamentos* allow rural populations to remain in a living environment similar to where they were before but require the creation of a new community, in which the migrants are the first stakeholders.

After buying a large piece of land, the dam company plans and organizes the space to accommodate the *atingidos*. The original property is then divided into various plots. Each family receives a house and the accompanying piece of land. The plots of land are equitably distributed on the section. The area of land received by the families depends on various factors (area of purchased property, area of previous property, negotiations, etc.), but the plots within one *reassentamento* generally tend to be of similar size. In a rural *reassentamento* of southern Brazil (Barracão/RS of the Machadinho Dam), the plots are 17 ha. They include an area for agricultural production, an environmental preservation zone and the family's living space, with a house, a garden often including a vegetable patch, a garage, and a barn to store agricultural equipment.

In addition to a house and a piece of land, the *reassentamentos* benefit from the presence of community facilities. The type of infrastructure provided depends on the company's willingness to invest, negotiations with the group, the number of families, and other criteria, such as location and cost. To facilitate circulation between the houses, the company makes dirt roads. Although previously optional, electricity is now almost systematically provided in collective *reassentamentos*. Among the facilities encountered in the *reassentamentos*, we found: an agricultural shed to help with the families' production, a community center, churches for various faiths, a school, electricity and phone networks, and sports grounds (bocce or football). Some of these facilities influence more than others the establishment of a community life. Thus, we noted the importance of schools that allow families to get to know each other through their children but also the creation of shared "moments" at drop-off or pickup time.

The houses received by the families are all identical to start with, but get modified according to the *reassentamentos*' needs and tastes. The houses visited in the south of the country, in Barracão/RS, consist of a large kitchen/dining/living area, two bedrooms, and a bathroom. The houses are of simple construction, and the families seem satisfied. Improvements made to the garden and the house are a sign of a good adaptation to the place, and we saw several houses with a fertile vegetable patch and a colorful garden.

This compensation option is, however, not the most widespread. *Atingidos* must ask for it, negotiate, and wait. It is more common in the south of the country than in the north, but the option nevertheless exists for dams built in the Amazon. In the south, it is often favored by people who have previously participated in struggles and demands through social movements. In the case of Belo Monte, it is difficult for families to be granted the option to move to a *reassentamento*. The company is purposely showing a wait-and-see attitude to propose and implement a *reassentamento*. The possibility is only beginning to emerge after more than 3 years of construction and 2 years of family migration following injunctions by IBAMA. Norte Energia is finally starting to implement the project of a collective rural *reassentamento,* but the families who may have wanted this type of compensation have already migrated and therefore changed their way of life for 4 years.

The *reassentamentos* also play an important role in the destination municipalities by revitalizing the rural areas, in particular through the building of the facilities mentioned previously. The arrival of new families, the building of a school, the creation of a football team are all elements that fuel a new dynamism in rural areas, with population movements and new services and businesses. The *reassentamentos* occupy an important place in the rural areas of the municipalities that receive them and become part of the local territorial rationale, modifying it in various ways: mobility and circulation, territorial development, habitat and density, economic activities, stakeholders, etc.

Urban *reassentamentos* are a rarer solution in the landscape of the compensation options for the dam-affected families. We had the chance to visit an urban *reassentamento* of Altamira during the construction of the Belo Monte Dam and found that the *atingidos* tend to use the space differently in an urban *reassentamento* and that the building of a community around the group of families in the urban setting is also quite different.

Brazilian hydroelectric plants are generally built in rural areas, far from urban centers. A few urban areas are therefore affected by the construction of dams. This was nevertheless the case for the village of Itá, which was destroyed in the 1990s by the construction of a hydroelectric plant, causing the displacement of some 1000 people and forcing Eletrosul to rebuild a town a few hundred meters away, for a projected population of 2500. Another important example is the Sobradinho Dam which involved the construction of four urban centers for the *atingidos* (Sigaud 1988). More recently, the construction of the Jirau plant on the Madeira River, in the state of Rondônia, resulted in the creation of an urban *reassentamentos*. The company responsible for the dam built a new town (Nova Mutum) theoretically capable of accommodating between 5000 and 7000 people, dam-affected families who previously lived on nearby lands but also the employees working on the construction site. The mixed nature of the inhabitants could potentially pose a problem. A similar model was implemented for the Belo Monte Dam but separating the employees from the *atingidos*. The size of the project is also different as five collective urban *reassentamentos* were created for Belo Monte.

The first problem posed by urban *reassentamentos* for the companies is their setting up. We see two different models. The first consists in the replacement of one town by another nearby. It involves the building of a new city, an urban space created ex nihilo in a rural area. The second model, which we will focus on, is more complex for the company: It involves the extension of an existing city by creating new urban districts, most often in the city's periphery. This option poses a land problem because the purchase of properties near a city is difficult and expensive, while the first model only requires the purchase of land in rural areas.

The spatial organization of the urban model is very different from the rural one because the population density is much higher. Each family only gets offered a house and small garden. There are no separations between houses to start with, but fences are often the first additional element built by families. The district plan is quadrangular with paved streets at right angles, and houses are arranged geometrically along the traffic axes. Some "quadras" are left completely or partially empty to allow for the installation of future infrastructure, which breaks a little the monotony of the model. The houses are built with simple materials and are often small, with two bedrooms and a shower room. The building materials used are so basic that it can sometimes cause problem, as in Altamira where families promptly complained after discovering cracks in the walls only a few months after moving in.

The creation of a new district in an already large city proves difficulty in particular because of the lack of available land and its cost. It is also important to ensure suitable links between the new neighborhood and the city center so as not to isolate the families. When presenting its urban *reassentamento* projects, Norte Energia proposed to place them at a maximum distance of 2 kms from the families' previous places of residence (Eletrobrás, Eletronorte, MME 2009). In practice, the cost, land availability, and urban morphology meant that this could not be achieved. Thus, the centers of the five new neighborhoods built until 2017 are located about 5 kms from Altamira's town center. The integration of the new neighborhood also depends on its relations with the neighboring districts. In the case of Altamira, the neighborhoods adjacent to the *reassentamentos* have developed rapidly, spontaneously, and in an unorganized way, taking advantage in particular of the economic dynamism, the collective facilities, and the financial windfall of the migrant families.

Collective urban *reassentamentos* (RUC) differ from rural ones by the fact that they almost systematically benefit from community facilities. The infrastructure construction depends on the level of investment from the company, the number of families resettled and their needs, and land availability. They may include a school, dispensary, police station, place of worship, sports field, park, community center. Network infrastructure is also put in place such as water, electricity, sewage, Internet, garbage collection, and public transport. Once completed, the new neighborhood often shows a clear improvement compared to the families' old place of life. The implementation of these facilities is, however, not standard procedure, and the time frame is quite variable because their construction depends on the finance and the will of the company. To meet the needs and pressures, the dam company opts to build houses first, before taking care of the rest. It can then fall behind or completely cancel infrastructure constructions, making it harder for families to settle in the new neighborhood.

Urban and rural *reassentamentos* undergo rapid changes after the arrival of migrant families. From the very beginning, families invest in changes in the structure of their houses and in their grounds. Protecting their house being essential for urban *atingidos*, the first modification they do is often to build a fence, sometimes with proper walls, high and private. They then sometimes choose to extend their house by building another room or shower room to get more space for the family. The garden, in front of or behind the house, also often gets transformed by creating a small vegetable patch or building a storage shed. In Altamira, families recovered materials from their old house to reuse in the new one. The other significant change is the creation of businesses and services, which can be associated with a desire to settle and adapt.

While the urban structure seems adapted to receive migrants, problems quickly appear as much for the populations as for the local stakeholders. For example, the question of infrastructure proves problematic in the medium and long term. The management of the facilities built by the company, and of the neighborhood as a whole, is then retroceded to the municipality which must cover the operating costs. A rapid degradation of services is sometimes observed (Nova Mutum), prompting people to move or migrate again. Another issue is urban violence. A true plague of Brazil, it seems exacerbated in these new neighborhoods causing fear and instability for the families. In 2017, Altamira became the most violent city in Brazil, with the highest homicide rate (IPEA 2017). In addition, an urban life framed by the "system" can also prove difficult for families. They must adapt to a more structured way of life, governed by bureaucracy and administration. Take the frequent example of the electricity bills that families must now pay when previously they simply occupied urban spaces illegally. Families often protest against the bills' exorbitant amounts and the impossibility of paying them. The mismatch between the new urban setting and the *atingidos*' old way of life is often reported.

The cultural difference is problematic to such an extent that RUC has been considered unsuitable by some researchers (Stolerman et al. 2014). In the case of Belo Monte, populations that are neither fit nor ready to change their lifestyle have difficulty settling in. For example, the lack of access to the Xingu River is highly detrimental for *atingidos* whose life depends on fishing. They cannot find alternatives and therefore seek to rebuild their past lifestyle. Some families travel 5 kms every day with their boat to work. A new *reassentamento* (Pedral) is nevertheless under construction in the south of the city to propose to these fishing families and the urban Indians a more adapted living environment in the shape of a *reassentamento* with access to the river. Rural families who come to live in an urban setting do not find their bearings and find it hard to adapt. We noticed in the five RUCs of Altamira that new homes had been abandoned, a phenomenon also reported for Nova Mutum (Stolerman et al. 2014).

Urban *reassentamentos* present advantages and disadvantages. Evolving very quickly, they are dynamic and allow families to adapt quickly. On the contrary, others do not manage to stay there. They also raise fundamental questions of management and urban structure, which should be resolved before the arrival of families (Fig. 2.5).

The district of Jatobá in Altamira helps to illustrate how a collective urban *reassentamento* is composed and evolves. Built in 2012, the new neighborhood covers an area of 88 ha and is located on the northern outskirts of the city, about three kilometers from the city center. Divided into 1249 plots of land, each with a minimum area of 300 m^2 and an average of 316.5 m^2, the district was designed to accommodate a total of 1250 families, according to the 2011 plans. The official accounts of Norte Energia from 2017 show a capacity of 1154 residences and an occupation of 1074 families. Each house has an area of 63 m^2 with two or three bedrooms. Made of bricks, houses were the object of many criticisms by the first inhabitants because of visible cracks. Each house has, at the front, a small grassy area and connections to the town services and meters, and a small garden at the back (Fig. 2.6).

Among the most important public service infrastructures, the district offers a dispensary, a police station, a population assistance center, and a newly opened school (2017), more than 2 years after the creation of the district. The installed networks are those of drinking water (large tanks at the top of the district), telephone and electricity and a sewerage system, which, according to the inhabitants, is not working properly. The roads are tar sealed despite this not being the case for all the streets of Altamira. As of a few months ago, urban buses also serve the city center. At the community level, the company Norte Energia has built a multisport area in the center of the *reassentamento*, a place of worship and a community center.

Having hosted for nearly 3 years, the headquarters of Norte Energia in Altamira, Jatobá, plays a somewhat symbolic and exemplary role. Moreover, it was the first urban *reassentamento* built for Belo Monte and it has used a lot for promotion and journalistic marketing accompanying the construction of the dam.

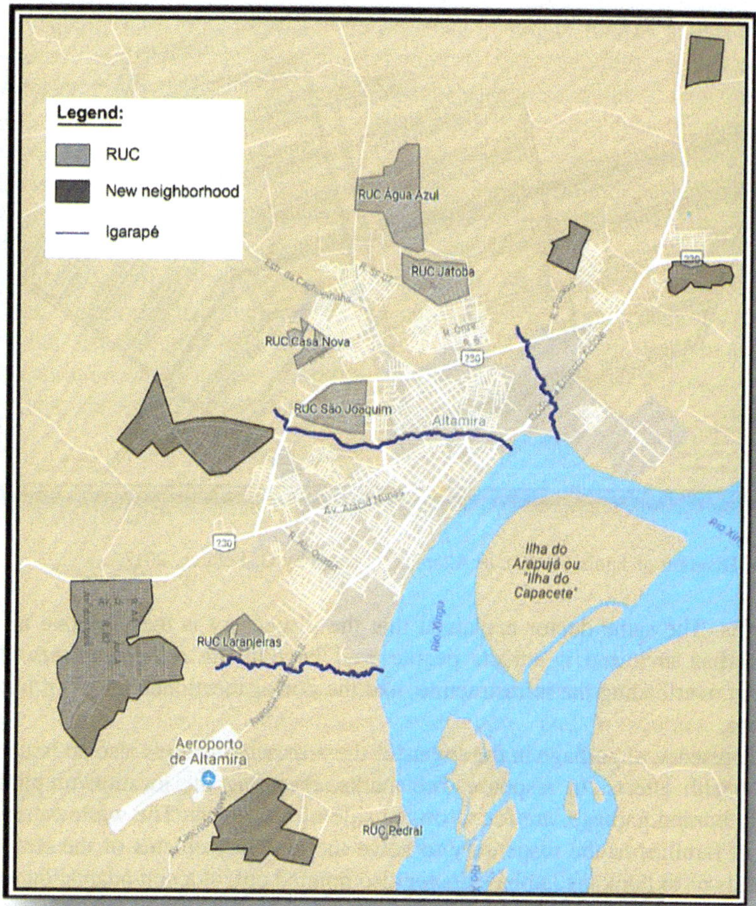

Fig. 2.5 Localization of RUCs and new neighborhood in the urban area of Altamira. Modified by G. Leturcq, 2017

In January 2014, the first inhabitant of the Jatobá neighborhood entered his new home and in the space of just under 1 year just under 1000 families had moved there. Since 2014, the neighborhood has evolved in terms of the collective facilities as well as the individual residences. A serious degradation of the infrastructure can be observed, in particular the water and sewerage systems for which the inhabitants do not stop protesting. The quality of the water distributed is problematic and the reason for complaints with the company due to digestive problems noted in particular at the local dispensary after interviews with the chief doctor (interview 29/05/2015). The water tanks are not sufficient, and every day tanker trucks have to come to supply the neighborhood. Some families have also dug wells on their wait before filling the reservoir and therefore to produce electricity. Families are therefore pressured to make a decision rapidly. In addition, the company has its own preferences in terms of the compensation options and strongly influences families to make them opt for the least expensive, simplest and fastest solution. *Atingidos* interviewed in the south and in the north of Brazil sometimes explain that they signed documents under the pressure of the companies' employees without really knowing what they would receive. This situation is also recounted by H. José da Rocha (Rocha 2013) who explains that companies try to control the negotiation process.

Families must make an individual choice, which can still be part of a collective framework in the case of a compensation in a collective *reassentamento*. The selection process can be assisted by a social movement or union, but at the very moment of selection only the family is responsible and signs the document. The choice is made by the head of the family, with the support of the other family members and with their own conditions and influences, such as their level of education. The simplest, most uninformed, and less educated families tend to see their choice more influenced by external factors and stakeholders. Outside help is relatively limited during the decision process as explained by the *atingidos* of Belo Monte, 95% of whom negotiated alone or with their families. The other stakeholders (neighbors, unions, friends, etc.) do not really weigh in at the decisive moment.

The choices offered evolve over time, both in the long term, because the options offered for mid-twentieth-century dams are not the same as today, and in the medium term, because the possibilities for a dam can change during the compensation procedure.

Dams of the early twentieth century had much less impact than today on the families living near the affected rivers, and we do not have accurate factual information on the compensation options available at the time.

From the 1970s, compensations became a public subject of discussion, particularly because of the claims made by the families affected by large dams (Itaipu, Sobradinho, and Tucuruí for example). The military had little regard for the families and the compensation they should receive, but their point of view changed with the protests and they had to adapt their proposals. Families in Sobradinho were "offered" three alternatives: the "lakefront," the Serra do Ramalho settlement project, and a migration out of the region with a paltry sum of money and a bus ticket (Sigaud 1988). These options, which were more proposed destinations than an actual compensation, became available when CHESF realized the strong desire of the families to stay as close as possible to the future reservoir (Sigaud 1988). This can be explained by the fact that the proximity to the river favors the flooded farming practiced by the local families.

For the Tucuruí Dam, the choice of compensation was also governed by the migration's final destination. Eletronorte decided that rural families should move to a rural *reassentamento*, urban families to a new urban quarter, and also offered a third option, monetary compensation, in order to encourage and help families to leave even before official relocations. In exchange for financial assistance, families left and signed a "surrender agreement," giving up their lands and properties.

This period of Brazil's history of dams is strongly marked by the military power which governed the country with authority and very limited freedom. Dam-affected populations were treated no better, quite the contrary, and received almost no consideration when choosing and receiving compensation for their losses. Threats, intimidations, and coercive methods against the population were reported (Sigaud 1988). The process of compensation and migration for the Sobradinho Dam has been compared to a territory military evacuation, thus highlighting the violence and the methods used.

The 1980s saw rebellions in various parts of the country. From Tucuruí right up to the banks of the Uruguay River, via Itaipu, social movements for fair compensation claims arose. Demonstrations, encampments, marches, blockades, and all other possible methods of social activism helped to make the voices of dam victims heard. These helped highlight the need to review the methodology for treating migrant families and to offer them more compensation options for their losses.

A first result appeared in the south in 1987 with a signed and ratified agreement (PRE-692/86, of October 7, 1986) between Eletrosul and MAB, formerly CRAB. This agreement had two components (Radaelli 2010). The first concerned the impossibility to construct a dam before compensations, and/or *reassentamentos* options had been offered to the populations affected by the Itá and Machadinho dams. The second component stated the necessity to offer a minimum of three options to affected families: land for land, i.e., a new piece of land, in better agrarian and structural conditions than the lost piece of land, in the southern states; financial compensation with the involvement of the affected family in the assessment of losses; and the guarantee that the affected families without lands can, if interested, participate in the projects of rural collective *reassentamentos* in the two southern states, with better agricultural lands and facilities than those flooded. This was a real breakthrough in recognizing the options available to victims. The agreement is broadly respected and serves as a working basis between southern families and dam construction companies.

The families' choices thus depend on the proposed options. In recent years, the many dams built often offer the same possibilities. As there is no standard or legislation on the subject, a prior agreement must be signed between the social actors involved, mainly the dam company and representatives of the population. The agreement establishes the rules for the proposed compensation and *reassentamentos*, first giving an outline, then all the details. It defines the beneficiaries and gives general guidelines. For example, the guidelines given in the TA for the Mauá Dam were: to avoid a rural exodus, to preserve local culture, to offer a better quality of life to the *atingidos*, to offer technical, economic, and social support, to encourage participation, etc. (UHE Mauá 2009). The document then lays down the criteria (owners, tenants, inhabitants, fishermen, etc.) and the compensation procedure to clarify the process and avoid debates and negotiations. Finally, it explains the possible compensation options offered to the *atingidos*.

	Ita	%	Machadinho	%	Barra Grande	%	Campos Novos	%	Foz do Chapecó	%	Total	%
Indemnização em dinheiro	3260	78,4	873	38,4	959	63,1	449	59,2	1200	73,0	6741	65,1
Carta de credito	458	11,0	796	35,0	249	16,4	167	22,0	314	19,1	1984	19,2
Reassentamento rural coletivo	370	8,9	200	8,8	193	12,7	81	10,7	44	2,7	888	8,6
Reassentamento em area remanescente	72	1,7	31	1,4	5	0,3	62	8,2	25	1,5	195	1,9
Outros			375	16,5	114	7,5			61	3,7	550	5,3
TOTAL	4160	100	2275	100	1520	100	759	100	1644	100	10358	100

Fig. 2.7 Table of compensations in some south dams. Data from Rocha 2013

was spread between businesses, case studies still in progress, and social rents. Still under construction, Norte Energia implemented a new concept of compensation structure: an urban district on the outskirts (Pedral) with direct access to the river and intended exclusively for urban Indians and local riverside communities finding it hard to adapt to the urban context of the RUC. This urban *reassentamento*, bordering the rural area, makes it possible to try to continue a little of the way of life of the *atingidos*.

The analysis of family options and choices demonstrates the importance of the context and the power play that exist during the construction of a dam. The framework of the initial agreement leads to an outline of the compensation and the future of the families. It is therefore important that families are well represented and negotiate precisely with the company to be offered the best compensation. Companies are maximizing their influence to encourage families to choose financial compensation, not just to control the total costs of compensation but also to take on a minimum of responsibility after construction.

2.3.2 *Negotiations*

Once the project is approved, the company in charge begins to carry out impact studies to have a detailed idea of the flooded area. Quickly, it becomes aware of the families who will have to leave their places of residence and those who will be able to stay. A socioeconomic register, listing all the families affected by the dam and its reservoir, is established. This is done well before the construction of the dam, which can be problematic as it sometimes has flaws, being incomplete with regard to the number of families, their assets, and land ownership. The register is a confidential document which is not disclosed to the public or to potentially affected families.

The register should be considered as a base, but companies often use it as a reference to pay the least possible compensation. Between the moment when the register is created and the moment of changing residence, there is a period can reach up to 5 years. For example, the register for Foz do Chapecó was completed in October 1999 and revised in July 2002, while the agreement with the families was not until November 2004 (Rocha 2013). This 2-year delay caused problems for the families. Local populations living close to the future dam suffer because they await, with uncertainty and impatience, the construction of the structure and the time of migration. This period of doubt is synonymous with stress and poses a planning problem, especially for the families' crops, but also for their life project. This period is also theoretically frozen in time and space, because it ignores migrations and developments, and anyone entering the area to be flooded during that time is not taken into account.

In an attempt to improve the process and to clarify the situation, a special decree was drawn on October 26, 2010, by the will of President Luiz Inácio Lula da Silva. Decree No. 7.342 aims to consolidate and clarify the Socio-Economic Register for Power Generation Works (Brazil 2010). The regulation and implementation of this register fall under the responsibility of the Strategic Center for Socio-Environmental Management, which in turn answers to the Ministry of Mines and Energy. The decree's objective is for the register to identify, describe, and publicly record the peoples affected by the power generation works, and it also involved the creation of the Interministerial Committee of Socioeconomic Registers to evaluate the established registers.

The register is compiled in several stages over a long period. Whether it is carried out according to the criteria of the decree or according to the criteria of the company, it requires various visits of the affected area and individual meetings with the families. It is necessary to gradually analyze at first the region, then each property, and finally all the local families. According to a document from the Ministry of Mines and Energy dating from 2012, the register, under the criteria of the decree, must include an individual register, systematized data and information, iconographic information, additional information, and an analytical description of socioeconomic and cultural relations. It is therefore a process that requires time and money and, if done right, can serve as a basis for negotiation and building a relationship of trust. In addition, families should have a single referent who accompanies them throughout the process, but the task is often divided between different companies which sometimes change during the process. Families must first deal with evaluation technicians for the land, house, resources, agricultural production, land morphology, social structure of the family, etc. Then, there are the agents (often social workers) responsible for compensation who come and see them. The families' assets' value is then calculated, and the company makes an offer of compensation, based on the criteria chosen by the family. From that moment, families start talking with the company's lawyers. Once the offer is presented, they can think about it for a limited time (60 days, for example, at Foz do Chapecó) and negotiate it and a second offer is then presented. This second offer is always the last because after that the company uses the judicial system to achieve its ends. We observe a distortion between the legislative framework that seems a good working basis and the reality that is found to be more complex and unbalanced.

The relationship between the migrant families and the company responsible for the dam can be quite complex. The company's strategy is to divide families and to treat them on a case-by-case basis rather than dealing with entire groups or communities. It multiplies the individual contacts and limits as much as possible group meetings (Locatelli 2014). The strategy focuses on facilitating and accelerating the compensation process and trying to pay as little as possible to the families. This "divide and conquer" method, as in the Latin proverb, is an attempt to weaken and use one's power to influence and control negotiation. The power is strongly asymmetrical in this relationship between, on the one hand, a company often with years of experience and with strong political, economic, and cultural powers and, on the other hand, farming families often in precarious situations and forced to change their place of residence. With this imbalanced relationship, families have very little chance to express themselves publicly and to talk about their ideas and feelings. During their exchanges with the company, they hope to finally be able to speak up. So here, we have a fertile ground for conflict, between the possible frustration of the families and the powerful company responsible for compensation. As a result of this situation, the companies once again put in place a strategy to facilitate the process, namely to prepare the families for the numerous upcoming visits, to make them calm down and to ensure the good reception of the employees through posters and letters. In this way, the company tries to reduce the tension in the families in advance because it knows that otherwise its work can become quite difficult.

In 2000, the WCD provided the way forward in terms of negotiations. It recommended that the company or government implement a "Free, Prior and Informed Consent" (PIC) (WCD 2000) which implies "a continuous, iterative process of communication and negotiation spanning the entire planning and project cycle." The commission insisted that the negotiation must be conducted together with all the concerned participants so that the final agreement is accepted by all. It also indicated that recognition of the families' lifestyle and quality of life is beneficial for the project and the compensation phase (WCD 2000). Finally, for families to feel truly free, the WCD recommended that they are involved in all the steps toward compensation and especially that they have resettlement options, which are not to be forced upon them.

Preparation can be an important factor to facilitate negotiations and balance the power relationship between the two stakeholders. Among the elements favoring preparation, access to information is paramount, but a prior knowledge of the dam installation process, the education of family members, a previous migration experience, etc., can also help. We can give the example of families, living by the Uruguay River, in the south of Brazil, who, before seeing an actual dam being constructed on their lands, received training from local social organizations and stakeholders such as churches (Catholic and Protestant), rural unions, and some intellectuals, as described by Sigaud in 1995. In the northeast of the country, the Pastoral Land Commission (CPT) also played a fundamental role in explaining to farmers the problems associated with dam construction. The objective of pre-dam training is to "inform the farmers in order to mobilize them against the dam construction project and to alert them to the 'risks' inherent to the company's project" (Sigaud 1995). Thus, while families are thinking of fighting the construction of a dam, they are also preparing to fight for their right to a fair compensation. The balance of power depends on various criteria for the families, including schooling. Families living on dam sites tend to be less educated because they live in rural areas. In Brazil, rural dwellers have on average 2 years less schooling, with an overall average of 8.3 years (Moderna 2017).

The mobilization of families in pre-training programs prepares pressure groups and social movements as we explained for the Itaipu Dam. The struggles in Itaipu and around the Uruguay River helped to create the national movement for the protection of populations affected by dams, the MAB. The formation of family groups is also important for negotiations because, as a group, families can claim compensation that the company may not necessarily want to offer. Collective resettlement areas (*reassentamentos*) in rural areas are the main compensation option that can be negotiated as a group. The MAB still favors this structure for the resettlement of families and works in this direction. Prior training of the group that demonstrates an interest in this type of compensation may force the company to commit to this path. This is obviously not systematic as the case of the Belo Monte Dam demonstrates. This compensation option was theoretically available to the dam-affected families, but *reassentamentos* were not created until 2015, 4 years after the start of the project, and were only then offered to families who had left their homes since 2011 and had been moved temporarily to Altamira. The option was finally offered by the company because, after official complaints and lawsuits, IBAMA ordered Norte Energia to implement a collective rural *reassentamento*. We see from this example that, even though the families negotiated as a group to have access to compensation, the company made them wait 4 years before it started buying land for this project. This was a way to try and make families lose patience, as they had to temporarily survive in precarious conditions and worry about their future residence. The only way the negotiations could progress in this case was because of the obligation imposed by IBAMA.

The most common means of pressure used by social movements is the encampment. The example of the Itaipu camp (Germani 2003) shows how the pressure caused by the encampment had significant effects on the negotiation, but it took a considerable amount of time. Today encampements are being been replaced by other methods, more selective but maybe more visible. Construction blockages, marches, and protests are other actions observed around the dam construction sites.

Group negotiation can be beneficial, but it does not always lead to results. As Locatelli explains (2014), dam companies favor individual contacts. Nevertheless, we can observe that some groups have more influence than others during negotiations. Still in the case of Belo Monte, it seems obvious that Indian groups can put on pressure like no other set of individuals can, be it the *ribeirinhos*, certain communities, urban populations, etc. Indians enjoy some protection or at least some attention from the company because it knows their messages are more widely disseminated than those of other families in the region. As explained by Filho (Filho 2005), Indians of Altamira have long been fighting against the construction of the Belo Monte Dam and their actions have benefited from a great repercussion abroad since 1989 and the "meeting of the indigenous peoples in Altamira."

Finally, there is an imbalance in the timing of compensation, with higher compensation values at the beginning of the project than at the end. There are various reasons for this. The first concerns the company's budget. Companies often present an initial dam construction budget that cannot be met because it is under the real cost, as was the case with Belo Monte. Thus, at the end of the construction phase, financial resources are lacking or at least become more restricted, and the company then tends to compensate less. The other reason is a strategy implemented, especially in the case of Foz do Chapecó to give a high value of compensation to the people living in the construction area to encourage the families of the floodplain to also leave their homes. But despite wanting to negotiate after agreeing to move, these families did not receive what was paid to the first migrant families, and the difference can be quite substantial.

We must therefore ask the question of the future of families affected by a dam but not compensated because of a late arrival, issues with the register or other reasons. The struggle for recognition and negotiations is the key to having some legitimacy and trying to get compensation. Legal actions sometimes have positive results (Machadinho Dam) because justice can favor abandoned families struggling to rebuild their lives. The role of unions and social movements is fundamental to help families in their struggles and thus to rebalance the power between families and companies.

2.3.3 Violence

From the beginning of this book, we have been looking at the effects of dams on populations, in other words at changes and transformations that may or may not be well received by the people. This creates a conflict scenario between various stakeholders and engenders violent situations. Some violent consequences will be highlighted here by focusing on the dichotomy between symbolic and physical violence explained by Bourdieu (2001). For this author, symbolic violence requires and engenders the participation of the dominated in their own submission, that is to say a kind of coercion resulting from an involuntary and unconscious acceptance. Physical violence, more traditional in its approach, is, according to Bourdieu, a process of domination applied to those who do not obey spontaneously. Physical violence is observed empirically and produces a short-lived obedience, while symbolic violence has lasting effects (Landry 2006).

From an external perspective, the whole dam construction process seems like an unstoppable machine. Families living in the area are likely to feel helpless faced with an omnipotent system. The violence of the process can be likened to the structural violence caused by political and economic systems described by Galtung (1969). The construction of hydroelectric structures is part of these, and the state enterprise association weighs heavily during meetings with the local population. The dominance of one group over another is obvious and violence follows.

Psychological violence can endure in the long term. The families living around the site of the dam experience it at different levels and at different times. In general, psychological violence is insidious and demonstrates a desire to control individuals emotionally. A stakeholder who masters the psychological mechanisms can use them against another to manipulate, intimidate, and/or threaten. *Atingidos* are the victims of these behaviors. Even before the actual construction, local populations undergo a latent violence for lack of decision and information. The construction of a dam comes from the central power, "from above," and the local population hardly intervenes in the decision-making process. Information is scarce on the execution of the work, and the population remains in standby. Doubts arise and questions accumulate, making the waiting period agonizing. People wonder if the dam will in fact be built if they will have to migrate, if life after the dam can be better, etc. Anxiety grows with time. There are more rumors than there is precise information. Families often end up deciding to leave their home early. This allows them to migrate freely, therefore avoiding the period of stress caused by negotiations and dismemberment of the community, but it also means that they are not entitled to compensation.

The negotiation period is also a time of strong psychological pressure. Families are not used to making this kind of decision, which will change the course of their lives, and to carrying out negotiations. Pressed for time, they must decide on compensation and options. The negotiations are limited, with only two rounds of negotiation between the family and the company's representatives. If the negotiations do not result in a satisfactory outcome, the family enters a legal process that makes the situation even worse and more complex. Families also face the contempt of negotiators who are often more informed and educated, as seen in the collective meetings for Belo Monte. Time is a factor of the anxiety because as the end of construction draws closer, stress increases and the actual changes become imminent.

Dam construction is characterized by losses for families: loss of residence, community, property, links with the area, etc. Whether material or immaterial, losses are difficult to replace for families. The loss of the proximity to the river cannot be overlooked because it is the first element mentioned by the families affected by the Machadinho Dam (Leturcq 2010). The absence of the watercourse is psychological and sentimental. It marked the lives of the people and was part of their everyday life. Some elements also disappear with the rising waters such as places of worship, cemeteries, natural landscapes, etc. This all weighs emotionally and psychologically for the *atingidos*. Uprooting can lead to depression and suicide. In addition to a case encountered in the south, research and a documentary made on the effects of the Itaparica Dam are highly instructive about the mental health of the municipality of Itacuruba (Pernambuco), more than 25 years after construction (Cribari 2014). The suicide rate of this flooded municipality is ten times higher than the national average, and one reason may be the losses suffered by the local population. In the face of profound transformations, some people react very negatively and cannot cope, thus demonstrating the psychological violence experienced by families forced to migrate.

In contrast to symbolic violence, physical violence is tangible and measurable. It can be described by facts and evidence. It physically marks the actors involved, whether dominant or dominated. It can be more or less legitimized by various means, such as public authority or the legislative framework. More concrete, it is also more "sensational" and attracts more media attention, which can help carrying a message.

Economic violence, which can observe around dams, consists of withdrawing the means of production or doing work. The victims then become even more dependent. Forced migration is a way of taking away the means of production of small farmers by removing their land. Small farmers depend on the fields, not only for the sale of their produce, but mainly for their own consumption. They rely on the production of their piece of land. Thus, the initial resettlement assistance offered by some companies is fundamental to helping the families settle more quickly into their new living environment. These aids provide stability and peace of mind for migrant families to resume their lives and economic activities.

Administrative violence comes from public and political institutions. We mainly associate this type of violence with the military regime during which local populations were not at all considered during the construction of dams. They were perceived as a hindrance to development and as an environmental problem for dam construction by Eletrobras (Santos and Andrade 1988). Sigaud (1988) explains that, for the Sobradinho Dam, the lack of consideration for local people was so important, and the situation so chaotic, that when the waters rose, families panicked because they had not yet migrated. Some people committed suicide, and homeless families walked along the roads. The violence of the families' relocation process was much more intense during the military period.

Despite the return of democracy, state-sponsored violence continues through police interventions. This is considered legitimate violence, in the words of Weber (1919), in order to protect the interests of the government and the private company building the dam. There may be direct confrontations between the victims of the dam and the police who often protect the sites. The encampment of the Foz do Chapecó Dam was dismantled after 18 months in 2007 by a police intervention (Leturcq 2011). It should be noted here the insidious link between the police and the dam company which finances the local police in order to improve infrastructure and increase the police force. The National Intervention Force has also recently been operating to prevent the invasion of the São Manoel Dam project by Indian populations (Freire 2017). Police violence is often denounced by the social movement MAB, as was the case for the dam of Baixo Iguaçu in the state of Paraná (MAB 2016).

Judicial violence can also be reported. The judicial system frequently intervenes in favor of dam construction companies. Both for the declaration of public interest and for the legal repression of social movements, justice, often local, tends to support private entities. Due to unsuccessful negotiations, an *atingido* from the Foz do Chapecó Dam was forced by justice to accept compensation with a deadline of 5 days to leave his home (Leturcq 2010). For the same dam, a judge defended the interests of the company by condemning the leaders of the MAB social movement who encouraged the blockage of the site. The sanction, mainly financial, was effective in frightening and slowing down the protesters. The leaders of social movements are sentenced with the worst convictions, introducing a "criminalization" decried by social organizations such as the National Movement for Human Rights (MNDH 2006).

The police can also sometimes respond with violence in conflicts involving construction workers. The Commission on Human and Minority Rights denounced police violence at the Belo Monte Dam site among other significant problems and violence in the region (2012). During the construction of the São Domingos Dam, an internal and destructive protest movement arose and the police became involved. The intervention, violent according to the workers, was then reported in the media (Campo Grande News 2011). Police violence is generally denounced in Brazil, and dam sites are no exception.

The employees of the dam sites can also cause violent problems, to varying degrees, during construction. Conflicts can indeed arise between workers about employment conditions including safety and wages. This is a controversial topic because the information is controlled by the company, but police interventions have been recorded on some sites to restore order. Following the destruction and burning of some facilities at the site of Foz do Chapecó, the police intervened to stop a riot from breaking out after a fight between an intoxicated worker and a guard (Leturcq 2010).

Finally, violence can spread throughout the region around the dams, mainly because of the financial windfall that comes into the hands of construction workers and the lack of supervision from the company, public policies, and the police. Often denounced by the local inhabitants, violence is a serious issue during the construction of dams of all sizes. It became even more of an issue in 2017 when the Institute for Research in Applied Economics (IPEA) published an Atlas of Violence in which Altamira, where the Belo Monte Dam is built, is presented as the most violent city in Brazil. With all the national media repeating the information, Altamira became for a few days the center of attention in the country in terms of violence and lack of safety. The homicide rate in the country is on average 32.5 per 100,000 inhabitants; in Altamira, the country's record, it is 105.2 (2015 data). The number of homicides was 114 in 2015 compared to 13 in 2000 (IBGE 2000). The Atlas points to the rapid growth and lack of foresight as potential reasons, elements obviously related to the construction of the imposing dam.

Violence manifests itself in many forms during the dam construction process in Brazil (Vainer 1996). Of varying intensity, it affects all the stakeholders involved and marks the territory, as seen in Altamira. Violence against dam-affected populations adds to the other difficulties that the families have to deal with. Finally, the lack of control and the role of the police are central factors in explaining the growth of violence and public safety loopholes around dams in Brazil.

2.4 Identity

2.4.1 Population Diversity

Dam-affected populations are very diverse due to the multitude of locations. Dams mark the whole Brazilian society, and over time an *atingido* identity is built. The struggles and disputes, somewhat organized, have indeed forged an identity which covers three categories of people affected by the constructions.

Can we talk about an *atingido* identity in itself? Do the victims of dams recognize themselves using this term? The term "*atingido*" is not new, but its association with the dam-affected populations started in the 1990s and gradually established itself from then on. At first, the term *atingido* referred to flood victims (Moracs 1996). In the 1970s and 1980s, dam-affected populations were named differently depending on the situation and the dam. Thus, they were called the "expropriated" for Itaipu or the "drowned" in the case of the Passo Real Dam (1973). The designation of the affected populations also depended on the stakeholders' position in the ongoing conflict. There are many examples of negative names used in the media or politicians' speeches, such as dam dwellers, flooded people, rioters, invaders. For Sobradinho, political and economic leaders named dam-affected families the "barranqueiros," expressing the very pejorative idea of a population that is poor, unqualified, underdeveloped, isolated, illiterate, self-sufficient, and limited in their social relations (Sigaud 1988).

Carlos Vainer asserts that the *atingido* status is consistent with a specific evolution: The individual goes from a passive situation, where he is affected by something, to an active situation where he reacts. He then fights for or against something: "The *atingido* is the one who does not want to be a victim" (Vainer 1998). For Silvio Coelho dos Santos, the identity is built from growing concerns about environmental and social issues (Santos 2003). The condition of the population changes right from the beginning of the installation of a dam in rural areas.

The *atingido* identity was first forged through the conflicts which took place on dam sites throughout the country in the 1970s and 1980s. It developed with the constitution of social movements which brought together different populations under the same banner. This new identity was then directly discussed and debated within the institutions that represented them, such as CRAB at the regional level in the 1980s. These discussions revolved around the identity and the definition of the *atingido* because conflicts and negotiations depend on a clear understanding of the term. The debates concerned the definitions for populations who are or are not flooded, relocated, land owners, etc.

Discussions continued during the formation of the national social movement (MAB). In the space of 2 years (1989–1991), the terminology used during major national meetings evolved from "*atingido* dam workers" to simply "*atingidos.*" The *atingido* identity was then affirmed at the national level, for the whole country, and the recognition of the movement was already sufficient enough not to need any more detail. Based on the research of Sandra Faillace, Lygia Sigaud states that a "we," a sense of belonging, grew gradually with the social conflicts. In the long run, conflicts encouraged "populations to assume the *atingido* identity or to reinforce it" (Sigaud 1995). People increasingly recognize themselves as specific victims of hydroelectric dams and victims of a certain injustice.

In parallel with the affirmation of identity, there is also a denial of association with others. *Atingidos* do not want to be considered landless. They are owners or sharecroppers who occupy and produce on land and who get driven out of their homes by an event. The timing issue must also be addressed because it seems that identity is particularly strong during the social struggle and negotiations but fades after the migration. This can be partly due to the loss of the presence and influence of the social movement (Leturcq 2010). The *atingido* identity varies depending on time and location but also presents different subcategories based on different criteria.

The populations that seem the most affected by the construction of hydroelectric plants are those that live in direct interaction with the watercourse. Fishermen and *ribeirinhos* are the people for whom the river is most important. We find fishermen all over Brazil, while *ribeirinhos* are mainly present in the north of the country. They both rely on the river as their primary resource for survival and economic activity. It also helps with mobility as riverboats are the main mode of transport. Finally, it is an integral part of the landscape in the families' living environment. It regulates their daily rhythm, serving as a biological clock and ecological regulator. Its presence is an integral part of the *ribeirinhos*' way of life.

These families, due to the importance of the river in their daily life, suffer more from migration and its associated negative effects. Forced migration generally causes a loss of connection with the watercourse because the new residences are very rarely by the river. Land speculation and the difficulty of finding land to buy on the riverbanks make it almost impossible for *ribeirinhos* to maintain their way of life. Companies, however, do try to give fishermen who have to move some sort of access to the river as their main economic activity relies on the watercourse.

Companies sometimes attempt to provide these people with a living environment similar to what they were used to. There is also a federal prosecutor in Altamira who is fighting for a guaranteed direct access to the river for the victims of the dam whose way of life depends on the watercourse. His fight is a first, and his success, even if difficult, would offer a real chance to families to successfully adapt after forced migration.

However, these populations are usually neglected, or even forgotten, by companies. Some people are accidentally left out of the victim registers due to sloppy research on the part of the company and because they live in isolated areas. Fishermen are also a category neglected by the company as the conditions of compensation do not really correspond to their way of life. It is also important to note that, apart from the recent exception of Tucuruí, fishermen living downstream of the dam are never considered as affected by the construction (Mérona et al. 2010).

Ribeirinhos and fishermen are the first victims of a dam construction. They are the people who suffer the most from the consequences, including forced migration which uproots them deeply.

Farmers are the most represented category in the total number of affected families. More numerous than fishermen, who have largely disappeared from the Brazilian rivers, they occupy riverbanks all over the country.

They cultivate lands with reduced agronomic qualities that are difficult to work on as they are located on slopes which prevent any mechanization. Most agricultural economies found along rivers are food crops, usually for self-consumption alone. Some farmers also produce for the local community or the municipality, and some, especially in Santa Catarina, may have a commercial production, such as tobacco or chicken. Farming practices are simple, adapted to the land, and found alongside local crafts. The productions are obviously different depending on the region, but the great majority are on small plots of land.

The relationship with the river is a lot less significant for farmers than for fishermen. Farmers occasionally use the river for fishing, transportation, and recreation. Farmers surveyed along the Uruguay River speak of its presence in the landscape as a soothing element but not often as a fundamental criterion for the families' survival. The river is a defining element of the everyday landscape whose absence after the migration is less prejudicial for farmers than for other families.

The farmers' integration with the rest of society is more evident. Less geographically isolated than *ribeirinhos* and fishermen, farmers living close to a river have more relations with the urban area and its stakeholders, including trade unions, traders, politicians. They are more mobile and more in contact with urban areas, which plays in their favor in terms of information and negotiations when a dam is constructed.

Finally, in regard to migration and later adaptation, farmers have more facilities than fishermen but have to meet complex requirements. Their first major adaptation relates to the cultivation of a new type of farmland very different from what they were used to. Frequently located on plains, these new plots require innovations that the farmers must acquire the skills and finance for. To be able to survive on the new lands, the farmers must begin to practice commercial mechanized farming on larger plots. The change of scale is important and is not necessarily a success. Farmers generally benefit from psychological, social, technical, and agricultural aids from the dam company. These help with the initial acclimatization and more concretely with the development of new techniques, such as learning how to use a tractor, the requests for agricultural loans, the implementation of irrigation.

Farmers are the most recognized *atingidos* by the dam company and usually benefit from help and assistance to resume life on a new land. The loss of the river is important but not insurmountable for the families' settlement into the new living space.

Urban populations affected by rising water levels are rarer. There are indeed a very few urban areas affected by the construction of a dam, mainly due to the sheer number of people that would be affected in a city where population density is high. Companies therefore plan their projects to avoid cities and the complications of families' resettlement as much as possible.

Families in urban areas commonly live in precarious conditions. They live in floodable areas which flood without the basic sanitary conditions being met. The known examples of Altamira are eloquent as families live in urban river branches, in makeshift wooden houses, built over water on stilts, without decent infrastructure such as health facilities, drinking water, electricity, streets. The change of residence for these families is usually associated with improved facilities.

Urban populations affected by a dam do not usually have very strong links with their community and their neighbors, which may change in the new environment. Neighborhoods affected by the rising waters are poor and highly disorganized. Once the migration is complete, families settle in a planned neighborhood with many collective facilities (dispensary, school, streets, plots of land, etc.) and in a community under construction.

From an economic point of view, the changes regarding income-generating activities are not as abrupt as for farmers. Depending on the area of resettlement, people can usually keep their jobs but may have to travel further or by using a different form of transport. Urban *atingidos* show a real economic dynamism with the creation of shops and services in the new virgin space.

Since the migration distance for urban *atingidos* is usually shorter, just a few kilometers away, families need to change their mobility habits. In Altamira, there has been an increase in the number of individual vehicles, which were purchased with the Belo Monte Dam compensation.

The relation with nature is different because for them it is not the first resource and it simply serves like a daily supplement as demonstrated with the very many transformations of the collective urban *reassentamentos*' gardens into small vegetable patches.

Overall, the adaptation needed for these families is simpler because they are faced with fewer changes, including financial ones. Nevertheless, these families have fewer possibilities for compensation and are left without choice for the final destination of the migration.

2.4.2 Social Movements

The identity of the *atingidos* is formed in the fight against dam construction and for the recognition of their rights as citizens. Conflicts emerged amidst a multitude of actions and stakeholders throughout the country during the 1970s and 1980s. They formed the basis for a national social movement born in the early 1990s and still effective today which stresses that "water and energy are not goods" (MAB) according to the 2018 slogan.

The first fights for the defense of dam-affected families took place almost simultaneously in all parts of the country. From Tucuruí to Itá, through Sobradinho, Itaparica, Machadinho, and Itaipu, families organized themselves with the help of other institutions and formed local and regional groups of resistance: CRAB (regional commission of dam *atingidos*) in the south, CAHTU (commission of the Tucuruí Dam *atingidos*), and CRABI (regional commission of the *atingidos* of the Iguaçu River).

In the 1980s, the repertoire of strategies for social movements was provided: demonstrations in front of headquarters (e.g., 5000 people in the streets of Erechim/RS in 1987); marches (between the flooded countryside and the town of Foz d'Iguaçu for example (Germani 2003); encampments; expulsion of construction workers from the sites; sabotage, etc. The number of actions helped in the recognition of the various movements which sought legitimacy to take part in the negotiations and be listened to.

Social movements' actions are highly publicized and help to convey otherwise unheard voices. Eletrobras, the national company responsible for dam projects, only began to empathize with the dam-affected people when the social movements started to better construct their speeches (Moraes 1996). CRAB played a key role as it began to turn to dialogue and no longer had a discourse built solely on a refusal. In the late 1980s, meetings between *atingidos* multiplied in the south, bringing together many participants and creating "mechanisms for vertical and horizontal links for various regional, municipal, and local authorities" (Moraes 1996). The demands, the organization, the representatives, and the whole movement became structured and credible in the eyes of the electricity sector's stakeholders.

External stakeholders help to establish and then legitimize social movements. The churches and trade unions of the regions around the dams participate in the mobilization and organization of the victims (Poli 1999). These institutions play a fundamental role in forming the basis of social organization. CPT, for example, supported the populations affected by the Sobradinho and Tucuruí dams on a daily basis during the 1980s, helping them to become aware of the current phenomenon and encouraging them to think about a way to confront the growing injustice. As proof of the evolution of CRAB, there was "an incorporation of the commission's speech (CRAB) by other stakeholders" (Sigaud 1989) and the movement relied on new support such as the Federation of Agricultural Workers, the Commission on Human Rights, the Advisory Center, Public Education (Warren and Reis 1989). Other social movements and unions joined the fight, with rapprochements with the Landless Workers' Movement. In addition, academics and iconic researchers such as C. Vainer began to study the issue and to become involved in the 1980s. More focused on defending the rights of the Indian populations, they joined the dam victims through these peoples. The academics' support was accompanied by support from national and international NGOs, particularly for the most notorious dams, such as Tucuruí and Itaipu. Finally, pressure from the World Bank, in the late twentieth century, pushed the leaders of the electricity sector to listen and hear the protectors of the environment and the movements of defense of the populations (Santos 2007). Overall, people started to receive more and more support and their union within a national social movement increasingly seemed like an obvious choice.

In 1989 and 1991, two popular national meetings allowed the creation of the National Movement of People Affected by Dams (MAB). Populations from all parts of the country came together with converging interests and similar struggles. The movement united around the slogan "Earth Yes, Dams No." It established on March 14 as the national day of the fight against dams and was organized from the local situations for a national action. The objective was to assert itself in the face of plans to build new dams in the country. MAB claims to play a central role in changing Brazilian energy policy. The specificity of MAB, in comparison with other social movements, is an "anticipation of the dam construction process" (Poli 1999). The social movement is hierarchical with grassroots groups in communities and neighborhoods, overseen by a local, state, and national coordination.

The 1990s saw the movement change due to the national privatization context, notably of dams. The fight, already complex when dealing with public authorities, became even more difficult when faced with private companies. MAB had to assert itself even more in order to have its voice heard, which meant strengthening national alliances, particularly within Via Campesina, and expanding internationally. In 1997, it launched the first international meeting of dam-affected people in which twenty countries took part. MAB then participated in the World Commission on Dams in 2000 (Vieira 2001) and went to various World Forums. The environmental conflicts also allowed MAB to have more space for dialogue, as explained by Acselrad (2010), who placed MAB among the movements of militant ecology, saying that it "accused the economy of the sector for making its investments profitable by expropriating people from their environments" (Acselrad 2010). The movement continues today through the action of young activists who follow the fighting path of their elders and benefit from their experience (Fraga 2013). We also observed the strong participation of young people in Altamira in the fight against Belo Monte and even more for the improvement of the *atingidos*' living conditions. The link with the academic world is now even stronger, especially since the opening of special places for *atingidos* and their leaders in some universities of the country (Pereira and Perruso 2011).

The actions of MAB in the field of conflicts are diverse, and to better understand them we can highlight two areas of resistance. The first consists in the creation and the coordination of the Alpestre encampment, in 2005, in front of the future construction site of the Foz do Chapecó Dam. The establishment of the camp in a region that had known the action of MAB for several decades had several objectives at different levels. It aimed first to make local families aware of the future location of the dam in the landscape. The choice of location is important and symbolic because it allows to make an anticipated contrast with the future dam that will be built at the exact place of the camp. The encampment also had regional and national goals, as explained by a leader of MAB during a radio interview: "to denounce the problems that exist with these structures and alert the whole society so that another serious problem, which in addition to being an environmental crime is also a social crime, does not occur again in a nearby region" (Leturcq 2010). MAB mobilized the local population to occupy and organize the camp. For nearly 2 years, a small group (about fifteen families in July 2006) took turns to maintain the camp. The organization was not easy every day: The camp and MAB sometimes had to confront the police and the justice, which in a joint action managed to dissolve it in December 2006 (Leturcq 2010). According to MAB data, almost 1500 families went through the camp and overall it did fulfill its objectives of spreading warning messages, both locally and nationally, against dams, of popular mobilization and of fair claims.

For the most iconic dam of the last decade, Belo Monte, the action of MAB is a bit different. MAB does not have an official hall or office in Altamira to meet local populations and stakeholders. In a joint interview with MAB activists (some locals and others from further away), in a house in a neighborhood soon to be flooded, in February 2014, we became aware of the local reality of the struggle. MAB action is not visible at the municipal and regional levels but mainly at the community and neighborhood levels. The impression of a mighty power mentioned by activists talking about the company (Norte Energia) and its influence on the debate explains why MAB is not more present during discussions or more generally in the city of Altamira. Since Norte Energia uses all possible means for its publicity, MAB can only fight at a level other than that of the communities. Another reason proposed to explain the weak presence of the movement in Altamira is the precarious life of the dam-affected families. Having almost nothing to lose with the rising of the waters, MAB's anti-dam discourse fails to get through already neglected populations. In contrast, the message of modernity and development associated with the dam has more impact. MAB receives some local support (Catholic Church, justice) and occasionally participates in resistance and struggle events, but divergences appear with other social movements (Fleury 2013). There is really a discrepancy between the expectation we can have regarding the action of a social movement of national scale and the reality discovered and described by the actual activists. Finally, MAB is more present in the daily life of the populations and in the RUC after migration.

The action of MAB is sometimes visible after the migration. In Altamira, we noticed several specific actions that illustrate this fact: entertainment in the RUCs, Jatobá, for example, and negotiations with the company to improve the living environment of the families in the RUCs. The local actions of MAB in this phase contrast with the grievances of several *atingidos* in the south in 2007, 5 years after their migration. The populations complained about their abandonment by MAB, its disappearance from the "world" of the *atingidos* after the migration, in the collective rural *reassentamentos* as was the case for the families who resettled individually. The absence of MAB in the rural areas of the south outside the conflict zones for the construction of a dam can be explained by an urban activism more practical and effective than dispersed rural actions.

MAB's actions are diverse and multifaceted in the sense that it operates in virtually every state in the country. Nevertheless, its presence is more marked for dam projects of a certain notoriety. Actions can also depend on the national political context. At the dawn of the twenty-first century, the left wing came to power in Brazil through President Luiz Inácio Lula da Silva of the Workers' Party (PT). MAB and other social movements were hoping to see the birth of a progressive social project in the country which could have meant significant social improvements for the *atingidos*. As the changes were not necessarily those expected, the movement tried to put pressure on the government with some success (decree creating the socioeconomic register of *atingidos*). Nevertheless, MAB's voice remains rather limited and confined to the scenario of Brazilian social struggles.

The fight continues locally with the continuity of actions for dam projects. Nationally, MAB remains present and intends to have a place in the race for the 2018 presidential elections. The presence of ex-president Lula, probable candidate in 2018, at the opening of the national meeting of the movement in October 2017 is insightful into that matter. Finally, the struggle is increasingly being handled by the judiciary system and the regional and federal authorities, as we have seen in Altamira for Belo Monte but also in other cases such as for the Garabi-Panambi Dam which recently saw its environmental licenses blocked by judicial order.

2.4.3 Evolution and Barrageiros

The identity of the *atingidos* may change over time and depending on the circumstances. The populations of the collective *reassentamentos* create a new identity based on their experience as *atingidos*. Others get involved in the social struggle and become active members of MAB. In parallel, dam workers follow construction sites and develop a very different identity. There are a multitude of profiles in the world of dam construction.

The identity of the dam-affected populations can change over time, but some people become *atingidos* even before they become officially and practically affected by a dam. In the 1980s, populations received training on the impact of dams in southern Brazil. Some enthusiastically came on board and became so involved in the issue that they became members of the social movement and carried the emblem of the group. This engagement transforms a simple *atingido* into an activist. The actions of the movement help to federate and retain members. Huge marches (local or national) also have a power of attraction and consolidation and are often cited as key moments in their personal life and the reason for their engagement.

Families who migrate to go and live in a new space, a collective rural or urban *reassentamento*, must rebuild part of their identity because of new social relations and changes in their economic activity. People we met no longer recognized themselves as *atingidos* but as belonging to a new living space. In Altamira, families say they come from Jatobá, a new urban district, and do not necessarily refer to themselves as *atingidos* from Belo Monte. The construction of the new community is important, and activities help forge the new identity. For example, in the south of the country, Sunday sport activities have a positive impact on the families and help to unite the group. This can encourage people to become active members of the new collective structure. The more years pass, the more the *atingido* identity fades away and gives way to new identities.

The identity of the *atingidos* plays a fundamental role during the pre-negotiations with the company. Families may try to change their origin and identity in order to benefit from better treatment and compensation. This situation is described by Sigaud (1988) around the Sobradinho Dam where families pretended to be caatingueiros (inhabitants of the caatinga) in order to gain access to land near the future reservoir which simple farm families would not get. We see the same behavior in

Altamira with *ribeirinhos*' families trying to prove their Indian origins in order to receive special attention, more support and more favorable compensation.

Hydroelectric plant employees can also form a group, a kind of community, with a particular identity based on their work on the dam. *Barrageiros* describe themselves as dam workers, but other criteria help complement this definition. They are migrants, even nomads, moving from one construction site to the next, with some period of inactivity in between. Local workers on construction sites are also named *barrageiros,* but a clear differentiation exists based on mobility (França 2007). They often come from poor areas of Brazil, such as the northeast or the southeast. Too few scientific studies have been carried out on the subject, but some informal interviews have confirmed these origins. *Barrageiros* often move from site to site. For example, many Belo Monte workers came from the Jirau and Santo Antonio dams and are waiting for the start of construction of the Tapajos or Marabá dams. They can thus be considered without roots, always being outsiders at their place of work (França 2007).

The *barrageiro* identity dates back to the 1970s during the dam construction boom. At the time, they already formed a group of workers who trained and specialized themselves directly during their first professional job. They represented all specialties and professional activities (Quintella 2008), from the simplest, without any qualification, to the highest in terms of academic level. They specialized themselves in the construction of dams. The period of return to democracy and privatization of the sector (1980–1990) saw less dams being built, and they had to wait and/or work in other areas. The construction of new dams at the beginning of the twenty-first century saw their return. The first *barrageiros* are honored in the Itaipu Dam Museum, where four rooms are dedicated exclusively to the workers, with photos, stories, music, audio and video interviews, etc. This place of remembrance tries to recover, preserve, and share the memory of their activities.

Barrageiros have a group identity that is based on this unique link with dam construction sites as described by Quintela (2008). They have a certain pride and a particular cultural identity. The *barrageiros* are recognized and identified as such. Their group is active on the Internet, the WhatsApp application, and on social networks, as they keep abreast of changes in construction sites and job offers. The profile of the workers is mainly male: Around 85% of workers in the civil construction sector are men, but women are now increasingly present and have started filling roles that were formerly considered exclusively male (Noradi and Maccari 2007). They form a united group of workers who are proud of their work.

Culturally, they also play a part, especially from an external point of view, because they are associated with positive as well as negative impacts. In a big city like Altamira, *barrageiros* are mainly associated with money and an increased economic activity: They are consumers, and businesses abound in the city. For example, during the construction of the Belo Monte Dam, special nights were organized in nightclubs, offering special prices for the workers, with advertisements in town to attract them. The money that construction workers earn is an important windfall for the entire region, including businesses, services, and transportation. They are also regarded as a reason for negative and illegal activities, such as prostitution, drugs, and violence. The term *barrageiros* is therefore also pejorative, and *atingidos* get offended when they are called that name.

The *barrageiros* are the useful workforce of the companies, which put an attractive policy into place during contracts but, when these come to term, many *barrageiros* do not know what to do. They are faced with several options: staying in the area, waiting for a new job, or going back to their region of origin. They sometimes stay on site and therefore help to boost the area in population and economic activities.

References

Abramovay R (1998) Agriculture familiale et développement territorial au Brésil. Ruralia, p 3

Acselrad H (2010) Ambientalização das lutas sociais—o caso do movimento por justiça ambiental. Estudos Avançados 24(68):103–119

Almeida NJR (2016) Saberes da pesca em unidade de conservação: os pescadores da RDS Alcobaça no Lago da Usina Hidrelétrica de Tucuruí/PA. PhD Thesis, Universidade Federal do Pará, Belém

Alves J, Júnior T (2012) A migração do trabalho para o complexo hidrelétrico Madeira. Anais XIII Jornada do Trabalho. Centelha, Presidente Prudente, pp 150–166

Angeon V (2008) L'explicitation du rôle des relations sociales dans les mécanismes de développement territorial. Revue d'Économie Régionale & Urbaine 2:237–250

Arnauld de Sartre X (2006) Fronts pionniers d'Amazonie. CNRS, Paris, p 223

Becker BK (1982) Geopolitica da Amazonia: A Nova Fronteira de Recursos. Zahar, Rio de Janeiro

Benchimol S (1977) Amazônia: um pouco—antes e além—depois. Umberto Calderaro, Manaus

Bourdieu P (2001) Langage et pouvoir symbolique. Seuil, Paris, p 426

Brasil (2010) Decreto n° 7.342, de outubro de 2010. Diário Oficial da República Federativa do Brasil

Carvalho A (2006) Água sobre terra: lugar e territorialidade na implantação de grandes hidrelétricas. Master Dissertation, Universidade Federal do Rio Grande do Sul, Porto Alegre, p 157

Cavalcante FC (2005) O processo migratório na amazônia vinculado à mobilidade pelo trabalho—o caso da uhe de Tucuruí. Anais do X Encontro de Geógrafos da América Latina, Universidade de São Paulo, pp 3345–3360

CRAB—Comissão Regional dos Atingidos pelas Barragens (1988) Documento de accord entre Atingidos e Eletrosul. A Enchente do Uruguai 20:8

Cribari I (2014) De Profundis. Video, 21 min. https://vimeo.com/158411290

Dean W (1989) A luta pela borracha no Brasil. São Paulo, Nobel, p 286

Droulers M (2001) Brésil, une géohistoire. PUF, Paris, p 310

Duqué G (1984) A experiência de Sobradinho. Cadernos do Ceas, vol 91. Salvador, pp 30–38

Eletrobrás, Eletronorte, MME (2009) Relatório de Impacto Ambiental—RIMA. Aproveitamento Hidrelétrico de Belo Monte. http://eletrobras.com/pt/AreasdeAtuacao/geracao/belo_monte/Rima-AHE-Belo-Monte.pdf

EPE—Empresa de Pesquisa Energética (2010) Relatório de impacto ambiental UHE Teles Pires

Estrela ES (2007) Descompassos e dissonâncias em torno das Agrovilas de Serra do Ramalho-Ba. Anals II Encontro Ciências Sociais e Barragens, Salvador

Ferreira LPF (2009) Reassentamento dos atingidos por barragens: limites da competência regulatória da Aneel. Senatus: cadernos da Secretaria de Informação e Documentação, 7 (2):50–63

Filho AOS (org) (2005) Tenotã-Mõ: alertas sobre as consequências dos projetos hidrelétricos no Rio Xingu. International Rivers Network, São Paulo, p 344
Fleury LC (2013) Conflito ambiental e cosmopolíticas na amazônia brasileira: a construção da Usina hidrelétrica de Belo Monte em perspectiva. PhD Thesis, Universidade Federal do Rio Grande do Sul, Porto Alegre
Flexa CE, Silva KCA, Cintra IHA (2016) Pescadores artesanais a jusante da Usina Hidrelétrica de Tucuruí, Amazônia, Brasil. Boletim do Instituto de Pesca, São Paulo 42(1):221–235
Fraga GW (2013) Memórias de terras e de águas: o Movimento dos Atingidos por Barragens (MAB) no norte do Rio Grande do Sul através da História Oral. Annals XXVII Simpósio Nacional de História, Natal
França GB (2007) Barragens e barrageiros: um estudo sobre os processos de construção de identidades coletivas em uma área de empreendimentos hidrelétricos. Master Dissertation, Universidade Federal do Parana, Curitiba
Freire W (2017) Força Nacional ocupa canteiro de obras da UHE São Manoel. Canal Energia. https://www.canalenergia.com.br/noticias/53038164/forca-nacional-ocupa-canteiro-de-obras-da-uhe-sao-manuel
Furtado C (1970) Formação Econômica do Brasil. Companhia Editora Nacional, São Paulo
Galtung J (1969) Violence, peace and peace research. J Peace Res 6(3):167–191
Germani GI (2003) Expropriados terra e água: o conflito de Itaipu, 2e edn. EDUFBA-ULBRA, Salvador, p 266
Gohn MDG (2003) Movimentos sociais no início do século XXI, 2e edn. Vozes, Petrópolis, p 143
IBGE—Instituto Brasileiro de Geografia e Estatística (2000) Censo Demográfico 2000. IBGE, Rio de Janeiro
IBGE—Instituto Brasileiro de Geografia e Estatística (2007) Brasil: 500 anos de povoamento. Rio de Janeiro, p 226
IBGE—Instituto Brasileiro de Geografia e Estatística (2010) Censo Demográfico 2010. IBGE, Rio de Janeiro
IBGE—Instituto Brasileiro de Geografia e Estatística (2016) Relação anual de informações sociais —RAIS
IPEA—Instituto de Pesquisa Economica Aplicada (2017) Atlas da Violência 2017. Rio de Janeiro. http://www.ipea.gov.br/atlasviolencia/download/2/2017
Landry JM (2006) La violence symbolique chez Bourdieu. Aspects sociologiques 13(1):85–92
Leturcq G (2010) Migrations forcées dans le Sud du Brésil: les *atingidos*. PhD Thesis, Université du Maine, Le Mans
Leturcq G (2011) Les barrages hydroélectriques au Brésil: contraintes et violences. Cahiers ALHIM, p 21. https://alhim.revues.org/3772?lang=fr
Leturcq G (2012) Accès et diffusion des données des barrages au Brésil. Annals of Sharing environmental information: issues of open environmental data in Latin America, Porto Alegre
Leturcq G (2016) Recomposition des territoires en Amazonie brésilienne, vol 26. Confins, online
Leturcq G, Santos DS, Fusieger A (2014) A usina hidrelétrica de Itapiranga: entre passado, realidade e expectativas. In: Rocha H, Pase HL, Locatelli C (org) Políticas públicas e hidrelétricas no Sul do Brasil. UFPel, Pelotas, pp 164–188
Lima LHM (2013) A construção da Usina Hidrelétrica e os primeiros sinais de segregação socioespacial em Ilha Solteira/SP. Revista Eletrônica da Associação dos Geógrafos Brasileiros, Seção Três Lagoas 18:142–161
Locatelli C (2014) Communicação e barragens: O poder da communicação das organizações e da mídia na implantação de hidrelétricas. Florianópolis, Insular, p 360
MAB—Movimento dos Atingidos por Barragens (2016) No PR, reintegração de posse na UHE Baixo Iguaçu termina com atingidos presos e feridos. MAB, São Paulo. http://www.mabnacional.org.br/noticia/tr-s-atingidos-por-barragens-s-presos-pela-pm-no-paran-0
Maia AG, Buainain AM (2015) O novo mapa da população rural brasileira. Confins, p 25. http://journals.openedition.org/confins/10548
Mendras H (1995) Les sociétés paysannes. Gallimard, Paris, p 368

Mérona B, Juras AA, Santos GM, Cintra IHA (2010) Os peixes e a pesca no baixo Rio Tocantins: vinte anos depois da UHE Tucuruí. Eletronorte, p 208
MNDH—Movimento Nacional de Direitos Humanos (2006) Criminalização dos Movimentos Sociais no Brasil, A Relatório de Casos Exemplares. MNDH, Brasilia, p 34
Moderna (2017) Anuário Brasileiro da Educação Básica. Editora Moderna and Todos Pela Educação, p 178
Moraes MSM (1996) O Movimento dos Atingidos pelas Barragens da Bacia do Rio Uruguai e a ação político-educativa dos mediadores, p 13
Moran EF (1981) Developing the Amazon. Indiana University Press, Bloomington
Nodari TMS, Maccari ECCP (2007) O trabalho da mulher na construção da Usina Hidrelétrica de Campos Novos. Annals I Encontro de Economia Catarinense, Rio do Sul, p 13
NE—Norte Energia (2017) 12 Relatório Semestrais UHE Belo Monte
Pereira TL, Perruso MA (2011) Construção da Trajetória de Militância do Movimento de Atingidos por Barragens. SBS Sociologia
Poli OL (1999) O movimento de Atingidos pelas barragens do Rio Uruguai—MAB. Leituras em movimentos sociais, Chapecó, pp 145–163
Quintella W (2008) A História das Maiores Obras do País e dos Homens que as Fizeram. São Paulo, Saraiva, p 422
Radaelli IM (2010) A trajetória do reassentamento Itá I, Mangueirinha—PR. Master Dissertation, Universidade Regional do Noroeste do Estado do Rio Grande do Sul—UNIJUÍ, Ijuí
Ribeiro GMC (2013) A construção da hidrelétrica Candonga e a desconstrução de modos de vida: memórias e histórias de trabalhadores em Nova Soberbo/MG. PhD Thesis, Universidade Federal de Uberlândia, Uberlândia
Rocha HJ (2013) Relações de poder na instalação de hidrelétricas. Ed. Universidade de Passo Fundo, Passo Fundo, p 348
Rocha GM, Pereira ICN (2003) População e recursos: crescimento populacional e o uso dos recursos hídricos na cidade de Tucuruí/PA. Encontro Transdiciplinar sobre população, Campinas, 3° Anais do Encontro Nacional sobre Migrações
Roche J (1954) Les migrations rurales dans le Rio Grande do Sul. Économies, Sociétés, Civilisations 9(4):481–504
Roche J (1959) La colonisation allemande et le Rio Grande do Sul. PhD Thesis, Université de Paris, IHEAL, p 703
Santos SC (2003) A Geração Hídrica de Eletricidade no Sul do Brasil e seus Impactos Sociais. Etnográfica 7:87–103
Santos SC (2007) Hidrelétricas e suas consequências socioambientais. In: Verdum R (org) Integração, Usinas Hidrelétricas e Impactos socioambientais. INESC, Brasília, pp 41–56
Santos LAO, Andrade LMM (org) (1988) As Hidrelétricas do Xingu e os Povos Indígenas. Comissão Pro-Índio de São Paulo, p 196
Sigaud L (1988) Efeitos sociais de grandes projetos hidrelétricos: as barragens de Sobradinho e Machadinho. In: Rosa LP (ed) Impactos de grandes projetos hidrelétricos e nucleares: aspectos econômicos, tecnologicos, ambientais e sociais. Marco Zero, São Paulo, pp 83–165
Sigaud L (1989) A política « social » do setor elétrico. Sociedade e Estado 4:55–71
Sigaud L (1995) Estado, energia elétrica e meio ambiente: o caso das grandes barragens. COPPE/Editora da UFRJ, Rio de Janeiro, p 184
Silva JP (2005) O processo de transformação sócioespacial do munípio de Itá/SC, após a instalação da usina hidrelétrica. Anais do X Encontro de Geógrafos da América Latina, Universidade de São Paulo, pp 14600–14615
Simmons AB (2013) Expliquer les migrations: une théorie à la croisé des chemins. In: Piché V (ed) Les theories de la migration. Ined, Paris, p 535
Stolerman P, Santos SC, Silva AA, Floriani N (2014) A implantação da Usina Hidrelétrica de Jirau no rio Madeira e os processos de desterritorialização em Rondônia. Terr@Plural, Ponta Grossa, 8(2):371–387
Théry H (2005) Le Brésil, 5e edn. Armand Colin, Paris, p 287

Turton D (2003) Conceptualising Forced Migration. Refugee Studies Centre Working Paper No. 12, p 17

UHE Mauá (2009) Termo de acordo para indenização aos atingidos da usina hidrelétrica de Mauá. p 21

UNDP—United Nations Development Programme (2009) Rapport mondial sur le développement humain 2009—Lever les barrières: Mobilité et développement humains. PNUD, New York, p 251

Vainer CB (1996) A violência como fator migratório: silêncios teóricos e evidências históricas. Travessia—Revista do Migrante 9(25):5–9

Vainer CB (1998) Deslocamentos Compulsórios, Restrições à Livre Circulação: elementos para um reconhecimento teórico da violência como fator migratório. Annals XI Encontro Nacional de Estudos Populacionais da ABEP, Caxambú, p 17

Van Hear N (1998) New Diasporas. UCL Press, London, p 314

Vieira FB (2001) Do confronto nos vales aos fóruns globais: um estudo de caso sobre a participação do Movimento de Atingidos por Barragens na Comissão Mundial de Barragens. Master Dissertation, Universidade Federal do Rio de Janeiro, Rio de Janeiro

Warren IS, Reis MJ (1989) O movimento dos atingidos pelas barragens do Uruguai. Cadernos do CEAS, Salvador 120:21–34

Weber M (1919) Le savant et le politique. Union Générale d'Éditions, Paris, p 186

WCD (2000) Dams and development: a new framework for decision-making. World Commission on Dams

Chapter 3
Comparison of South–North Effects

Abstract The third and final chapter of the book is a territorial comparison of the effects of dams, between the north and the south of Brazil. Based on the author's fieldwork and research, this part is divided into two: It first presents the most obvious similarities between the impacts of dams in the north and the south of the country, before describing the major differences that exist, especially with regard to the conflict and the social involvement of the dam-affected populations. The chapter presents many examples of dams in the north and south of the country and their various consequences.

Dams are present in all regions of Brazil but are especially prevalent in the south and southeast of the country. The southern region of Brazil comprises three states: Rio Grande do Sul, Santa Catarina, and Paraná. The aim of our work is to compare this region, which has a longer history in terms of hydroelectric plants, with the northern part of the country, where dams are more recent. The northern region includes the states of Acre, Amapá, Amazonas, Pará, Rondônia, Roraima, and Tocantins. These regions represent two extremities of the country, more than 3000 km apart.

Field research carried out in the north and south of the country justifies this comparison. The comparative analysis in fact makes perfect sense once on the ground, as the consequences observed in both regions appear to be the same. It contributes to a reflective debate, both personal and scientific, and feeds a real intellectual curiosity. In addition, there is the issue of national integration, especially for the effects of dams in the Amazon, which are the furthest from decision-making centers. Dams in the north are still in the news, while those of the south are more "discreet" in the media.

The wealth of information collected allows for comparisons. Research on Brazilian dams has been carried out in the two areas since 2005 through bibliographical inventories, field research, interviews, questionnaires, detailed analyses, and scientific exchanges with many colleagues.

Since 2005, I have been interested in the southern area and more particularly in the dams of the Uruguay River, along the natural border between the states of

G. Leturcq, *Dams in Brazil*, SpringerBriefs in Latin American Studies,
https://doi.org/10.1007/978-3-319-94628-3_3

Rio Grande do Sul and Santa Catarina. It was my main study area during my Ph.D. in geography (2005–2010). During this period, I completed four field trips to this region and other expeditions to other Brazilian regions to compare and obtain information. Field research first involves an exploration of the possibilities of research, followed by a long period of questionnaires and interviews, concluding with a phase to finalize the data and information before verification of various concrete elements. In total, 573 people were questioned for two particular dams: Machadinho and Foz do Chapecó (Leturcq 2010). Two semiclosed questionnaires were developed: the first for the victims of the Machadinho Dam, 7 years after the first displacements; and the second for the victims of Foz do Chapecó, which was under construction at the time. The *atingidos* of Machadinho were interrogated in seven *reassentamentos*, scattered several dozens of kms away from the dam. All questionnaires were then processed on SPSS and analyzed for the writing of the thesis. During the doctoral research, 57 semidirective interviews were carried out with actors and observers of the construction of dams in Brazil. The abundance of information gathered allowed us to complete the doctoral thesis in geography with a solid knowledge of the issues surrounding Brazilian dams and more particularly in the south of the country and in the Uruguay River basin, without forgetting ad hoc research in the northeast of the country, in the states of Bahia (Sobradinho dam) and Paraiba.

The specific research on dams in the northern region is more recent, with my integration into a research project (Fapesp) on the socio-environmental impacts of the Belo Monte Dam coordinated by Prof. Emilio F. Moran (MSU and Unicamp). This project involves various international researchers from different disciplines. Since joining the team with a postdoctoral fellowship, I have been able to deepen my knowledge of dams in the northern region. After updating my readings on the Amazon, the dams in the north, and the demographic history of the region, I started my fieldwork. Between 2014 and 2017, I took nine trips to Altamira and the Pará regions in order to acquire a general knowledge of the area and thus develop further the following steps: application of questionnaires to the population of Altamira, then to the victims of the Belo Monte Dam, analysis of the nearby rural area, and monitoring of the urban and social evolution of Altamira. In practice, research focused on the preparation and administration of a questionnaire among 500 households throughout the city of Altamira in 2014, a follow-up of families 1 year later, and a telephone questionnaire accepted by 115 people. In 2015, a specific survey for the *atingidos* was carried out in the new district of Jatobá, in the urban periphery of Altamira, with a follow-up with *ribeirinhos*' families in 2016 (35 identified families agreed to answer). Finally, in 2015, a large questionnaire was submitted to farmers on the Trans-Amazonian highway on 400 agricultural lots. I created and organized all these questionnaires and worked out the logistics of their applications on the ground during my various trips. In addition, interviews were conducted with more than 30 local stakeholders. A complete iconographic and historical research for Altamira illustrated the urban change occurring between 2014 and 2018. Finally, a search for digital data and various publications was conducted, thus forming a rich database on the subject.

These two long and comprehensive research projects, carried out in the south and north of Brazil, form the basis of the following comparative analysis. These are two non-exhaustive but extensively documented studies based on primary and secondary sources and built-up over more than 10 years.

With regard to electricity production, the two regions are comparable in terms of installed capacity. Production data can sometimes be controversial, because of the various sources, so we will rely on the information from the Information Bank of Production (BIG—ANEEL). With its 18 hydropower plants, the north has an installed capacity of 33,200 MW, a third of which may be provided by Belo Monte. The south has 54 hydroelectric plants and an installed capacity of 24,200 MW. The difference between the two regions is due to the age of the dams, as the first attempts of construction in the south date back to the end of the nineteenth century–early twentieth century. ANEEL indeed recorded dams dating from the early twentieth century in the south (Bracinho or Canastra), while in the north the oldest dates from December 1975 (Coaracy Nunes). In terms of electricity production and not installed capacity, the north and the south are almost equal with, respectively 25,900 and 24,200 MW produced. It is the actual production volume of the hydroelectric plants and not their potential that is inspected and constrained by the regulatory agency (ANEEL). The south produces 100% of the installed power, while the north only uses 78%. The difference is explained by the weather pattern of the northern region, which has wet and dry seasons. If the rainy season cannot fill the reservoir sufficiently to produce electricity during drought, the hydroelectric plant will stop working for a few months or weeks in the year. This is the most important weakness or criticism of dams in the Amazon by the electrical engineers. The map of the distribution of dams in Brazil (Cf. Fig. 2 in book front matter) shows a concentration of dams in some catchments in the south (Uruguay, Iguaçu, and Paranapanema for example), while in the north, dams tend to be more isolated. With regard to the electricity sector, the two regions are therefore different in terms of temporality, because the north is more recent, and in terms of diffusion, because the clustering of dams in the south contrasts with the isolation of their counterparts in the north. In terms of number of dams and potential, Chart 1 (Ministério de Minas e Energia 2007) indicates that the estimated and inventoried potential is much higher in the north in comparison with the rest of Brazil (Cf. Fig. 1 in book front matter). Given the estimated potential, the north is therefore the preferred location for Brazil's future dams and it is probably quite coveted by companies in the sector, as demonstrated by the construction of the Belo Monte Dam in 2011.

Based on our field experience and the knowledge gained over the past 10 years, we will try to compare some effects of hydroelectric plants on territories and societies in the north and south of the country.

This methodology was chosen to try to understand if differences exist due to the design of the dams or according to the territories and their characteristics. The comparison is a way of enriching the debate on the effects of dams and a way of better understanding the territorial differences between the south and the north. This approach has long been used in geography (nineteenth century with Reynaud 1984). According to J. Rodrigues, the comparison only becomes possible after certain common themes have been identified (Rodrigues 2012). For Claval (1972), comparisons are born of hypotheses about the relations which unite them (combinations), so we are looking for links between the north and the south in the effects of dams. We will obviously take precautions in the comparison, not having the possibility to be exhaustive and complete but trying to follow as well as possible the themes and methodology chosen.

3.1 Impact Similarities Between North and South

3.1.1 Landscape Transformation

Large-scale construction plans, such as those connected with power plants, always generate landscape disturbances, both in physical and in abstract ways. The notion of geographical landscape has already been discussed, but further analysis and individual examples are necessary to fully understand its relevance. Landscape transformation is a constant in areas affected by construction dams both in the north and in the south; to fully understand how these projects affected the landscape, it is necessary to tackle the role of society in the geographical environment. Landscape should be understood not only as a series of physical elements, but also as a mental, cultural, and cognitive construct, associated with memories, feelings, emotions, knowledge, secrets, dreams, projects, etc. (Bonin 2004). Each individual or group of people has their own perception and their own ways of interaction when it comes to their respective environments. As R. Verdum puts it, "Landscape Unit must be socially recognized for its forms, functions, structures and dynamics, and as an important part of the historical and cultural dimension of the place and the surrounding area" (Verdum 2011).

The construction site and the water reservoir itself are the two main elements that disturb landscapes and its relationship with the population. They change all the spatial logic of the countryside and its anthropic field, constituted by such elements as farming, trees, dirt roads, and isolated houses. Both in the south and in the north, the rural areas affected by the dams were also strongly connected to the riverbanks; the relationship between the population and the rivers also changed deeply.

Just as there are different kinds of affected landscapes, there are also different kinds of water reservoirs. Some are relatively small, and others are huge, occupying hundreds and hundreds of miles and, obviously, generating much greater environmental changes. Medium-sized water reservoirs include Itá (140 km^2) and Machadinho (79 km^2) on the Uruguay River, Curuá-Unã (121 km^2) in the Amazon basin and the Pitinga plant (80 km^2), which flows into the Jauaperi, a tributary to the Rio Negro. These reservoirs cannot be considered "small," as their influence over the landscape is not negligible. In the south, the largest reservoir is the Itaipu Dam, with 1400 km^2. The huge lake looks like a landlocked sea, and its limits are very wide apart. In the north, we find even larger reservoirs that deeply scar the landscape. Some of them are in the middle of the Amazon forest, such as the

Balbina plant, which extends over 3000 km^2. It is an impressive sight, with hundreds of dead trunks sticking out of water—a faithful image of the damage caused to the forest by this particular power plant. Other large reservoirs were created in the Amazon, such as Belo Monte (650 km^2) and Tucuruí (2800 km^2).

Another radical landscape change was caused by the Belo Monte construction in Volta Grande (Fearnside 2017), an area of about 100 km within the Xingu region, inhabited by 800 people along the villages of Ressaca, Ilha da Fazenda, and Garimpo de Galo, according to the information published by the IBGE in 2007. This area has been radically transformed, as the building of the dam entailed a change in the course of the river. The water flow has been hugely diminished, and the relationship between the population and the river has been inevitably transformed. As information about the construction and its consequences started to circulate, part of the population left, while others remained, despite the foreboding predictions. It remains to be seen how these people will survive in the future, with the loss of the river, and where will they take their livelihood from. It should also be noted that, on the following years, this area will be the site of the biggest open-air gold mine in Brazil: the Belo Sun mine, which has already finished the first stage of environmental licensing.

The construction of a power plant is such a radical element within a given territory that its consequences affect not only the physical aspects of a landscape, but also its imaginary representation. The building of the Belo Monte Dam has fed a huge amount of different points of view, both positive and negative, among local populations. As for the image of the Belo Monte project among Brazilians from other parts of the country, that is certainly a very interesting research topic—but one that must be dealt with in another occasion. The impact of dam construction on landscapes is an interesting element, a field still open to deeper research and a good starting point to analyze the impact of water power stations in Brazil (Fig. 3.1).

3.1.2 Migration Similarities

Anywhere a dam is built, migration takes place: Whole families are forced to move from their houses to a totally different area. Both in the south and in the north of Brazil, migration flows showcase some similar patterns. Whether in the south or in the north, migration is mandatory and compulsory. Even so, migrations can be classified into various ways, depending on their different levels of complexity, stages, period, types of populations, numbers, and reasons.

Population shifts start in the moment a new dam project is announced: At first, people move not from, but into the targeted area. Families and individuals are attracted by the new enterprise, in search of jobs and opportunities. Others try to pass by local residents in order to demand compensations. In the south, some families are known to have built impromptu wooden shackles near the Foz do Chapecó Dam just before construction begun. Elderly people, children, and other individuals seen as unfit to work were sent to occupy these houses; the idea was to pretend they inhabited the place, so the families could be granted money

Fig. 3.1 Landscape modification in Foz do Chapecó Dam. Alpestre/RS/Brazil, G. Leturcq, September 2007

compensations, even if just a little sum. The same pattern could be observed in Altamira: Several families moved in boats from different places to the Xingu riverbanks. Both in the north and in the south, families faced hard living conditions before moving to the vicinities of the construction sites; those conditions did not improve afterward. Apparently, their bid for compensation was mostly unsuccessful. The company had support from the police, and in many cases, the incoming families were forced to move. In both regions, the new economic opportunities attracted also people bent on illegal activates, like drug dealing or prostitution. Brothels were quickly built around the construction site in the Foz do Chapecó and near Belo Monte.

In Belo Monte, the company responsible for the dam is fully aware of these migrations and their negative impacts (Eletrobrás, Eletronorte, MME 2009). In the RIMA, Norte Energia is also aware of all problems related to housing, sanitation, health, transportation, etc. that always take place after construction begins. In both regions, north and south, companies have shown little or no preparation to deal with irregular migrants.

Things are quite different when it comes to the *barrageiros*. Companies in the north and the south have shown willingness to receive them, building special lodges and providing assistance and transportation for those workers. At the Foz do Chapecó Dam, a migrants' house employee stated the biggest problem they face when it comes to receiving new workers from the northeast and southeast: the cold.

Barrageiros coming from other parts of Brazil are not used to the low temperatures in the south and need to be given clothes and blankets. All over Brazil, building companies make little use of local labor. This probably happens because, in both areas, local workers are more used to agricultural jobs than construction.

As to the affected families, migrants go through very similar stages in the south and in the north. In many cases, as a family receives money compensation or is given a new house, it leaves its residence and proceeds to the new address. In some cases, however, things are not so simple. Some families move to the new house, come back to the old one, and then move again. Some are resettled in temporary abodes and fail to adapt.

In the southern dams of Machadinho and Itá, family associations spent several months demanding reparations from the company and the government and, meanwhile, remained without fixed residence, because their former proprieties had already been flooded. During that time they lived with relatives, waiting until they could start rebuilding their lives in a new home. In Belo Monte, families who chose to live in collective rural settlements had to wait more than 2 years until IBAMA had to intervene, making it possible for them to start the process of reinstallation in a new house. During this period, most families had to endure precarious living conditions in Altamira. In the Tucuruí Dam, several families have changed residencies at least twice after leaving their original proprieties. Reasons were diverse: Sometimes they faced unhealthy conditions; sometimes they sold their land to another family, etc. (Magalhães 2007). It should be noted that these cases are a minority, but they do exist.

3.1.3 Mobility Similarities

We understand mobility as the necessary movement of individuals or groups of people—for instance, from their homes to their jobs, places of leisure, secondary homes. As usual, we shall address everyday mobility, but also weekly and temporary forms of mobility that respond to specific logics. Mobility is affected by social and technological innovation and depends on several criteria, such as: spatial insertion, distance, accessibility, skills, activities. These criteria will be used to address mobility changes regarding populations affected by dams. Those changes take place before, during, and after the dam construction.

Generally, mobility is increased among populations affected by dam constructions. Studies in southern Brazil have shown that displaced families acquired new vehicles, especially cars and motorcycles, after their relocation (Leturcq 2010). It happens, mostly, because displaced people struggle to keep some sort of connection with places and individuals associated with their old lives. In Santa Catarina, we interviewed individuals who would travel up to 100 km to visit former neighbors and relatives, or simply to stay in touch with the areas they formerly inhabited. Thus, it becomes apparent that the old living grounds are not effaced from the lives of affected families; instead, their life space is increased, as the new and old homes become a sort of continuum. Even with the increased forms of mobility, contacts between dam victims and their relatives decreased after the relocation. Interviews conducted among the victims of the Machadinho Dam (65 families) revealed the following data: 36 families stated they used to visit relatives from one to five times per month before the dam construction, whereas 26 used to go on such visits ten times a month. After relocation, 53 families stated they now used to visit relatives one to five times per month, whereas only six families performed such social encounter from six to ten times a month. Geographical distance and life changing explain the decrease in visitation, but mobility is an attempt to solve the problem detected.

We shall now discuss the Jatobá resettlement, in Altamira, where we questioned 274 families affected by the Belo Monte Dam. Almost 60% of the respondents stated they had other relatives affected by the dam construction. The average number of relatives said to be affected by the dam is 5,4. Thus, there is a wide network of families affected by the dam. Proximity is an important element here: 77% of respondents said their relatives who were affected by the dam lived nearby, whereas 23% said such relatives lived far away. In the Jatobá neighborhood, 80% of all families claim to have relatives living in the same area; 64% claimed they visit relatives more than once a week; 23% visit once a week; in other words, 87% of respondents visit relatives at least once a week. As for the rest of the family, who do not live in Jatobá, the visits are less frequent. Thus, 40% visit more than once a week and 25% visit once a month. Thus, of course, distance makes visits less frequent and family ties are wider.

The location and the infrastructure of settlements have a strong influence on mobility, especially regarding the relations between rural and urban areas. In some cases, resettled families do not need wide means of mobility. Such is the case of Barracão, in Rio Grande do Sul. Families displaced by the Machadinho Dam now live in an area furnished with a local school, school buses, roads, a community center, a mobile dentist, and other resources. In settlements such as this one, most people simply walk around as they perform their daily chores. A very different situation is to be found in the Curitibanos settlement, located several miles away from town and displaying a poor infrastructure. Because of such circumstances, family members become dependent on different means of transportation.

Altamira faces similar problems. As said before, a new settlement is being built 10 km away from the city; once settled there, families will have to find new means of mobility in other to access services that are only available downtown. As the construction is still taking place, it remains to be seen if the new settlement will successfully address the needs of its population. Means of mobility depend on where relocation takes place, but the matter is also important during the dam construction itself, as the movement of populations to and from the construction site tends to become very strong and somewhat chaotic.

A strong and constant flow of people can be observed between Altamira and Belo Monte, for instance. The offices of Norte Energia and other companies involved in the dam construction are located in Altamira; therefore, the town became an activity center for all people involved in the project. Altamira also concentrates several kinds of services and entertainment options. As the construction site is 50 km away from the city, construction workers need wide means of mobility, in order to move from the decision center to their place of work; those circumstances create an intense flow of buses between the two sites. Another important aspect, specific to the north, is the traditional movements of the riverine population. The people who live by the Xingu riverbanks, taking their livelihood from fishing and small agriculture, are still connected to Altamira, where they must go from time to time in order to buy goods and search services. As the dam construction begun, the riverine population went more and more often to Altamira, now in search of information regarding the project and struggling to find more about how they would be affected in the future.

In the south, we observed a similar flow of people, although not as strong, around the construction of the Foz do Chapecó Dam. The city of Chapecó lies around 60 km from the dam, and the city is a regional center, in terms of infrastructure and transport (airport). The movement was less important than in Altamira because the economic impact of the dam construction was not as large as in the north. The western region of Santa Catarina has diversified economic activities, agroindustry being the main one. Thus, the construction of the dam intensified the mobility in the region, but did not create such a unique and specific population flow.

Urban mobility can also be affected by the construction of a dam. In Altamira, we observed an increased flow of cars and motorcycles on the city streets. One of the reasons for this is Altamira position as the region's epicenter. It concentrates all information and services. During the most chaotic phase of dam construction, a huge number of people had to come to Altamira to fulfill their basic needs. The relocation of affected families on the outskirts of town also responds for a significant increase in observed mobility. In Fig. 2.5, it can be observed how the city area has grown and also how the relocation of affected families influences population movement between the suburbs and the city center.

According to the National Association of Public Transport, more than 90% of public transport used in Brazil are buses and the amount of public vehicles tends to increase as the city grows. Until recently, Altamira did not have a considerable public transport system. Now, public transportation is available, but it is very hard to get any kind of information about bus routes and schedules. Therefore, for most of the population, public transportation remains invisible. It seems obvious that urban planners in Altamira will soon be faced with the urgent need to improve those services or make them more accessible to its population, especially now that the city is experiencing a strong horizontal growth.

A similar phenomenon took place around the dam built between the Brazilian town of Foz do Iguaçu and Paraguayan town of Ciudad del Este. The integration of the plant between those urban areas was a complicated affair—among other things, because we are dealing with an international border. As Adelita Araujo de Souza (2011) explains, the construction of the dam affected directly the development of both cities. During construction, there was a great increase in population movement and economic exchanges. Souza explains that "the urban motivations of this era, both in Brazil and Paraguay, were linked to the objectives of building the ItaipuBinational plant" (2011) and that "the urban future of the two cities was directly decided by the actors responsible for the construction of the plant." Today, we can say that the dam and its associated activities (technological development, university campus integration, and tourist center) are integrated into a metropolitan area that encompasses both cities.

The Santo Antonio and Jirau power plants, in the north, near Porto Velho city, had great importance in terms of population movements and mobility in an even more specific way. As shown by Cavalcante et al., population movement increased so much that a new territorial management plan had to be developed to integrate population dynamics, economic activities, and land use issues (2011). Because of the dams, the city decided to hire the company responsible for the Santo Antonio plant to carry out a new Urban Mobility Plan for Porto Velho (Santo Antônio Energia 2015).

3.2 Differences from Impacts in the South and North

3.2.1 Differences in the Social Conflict

Since the 1980s, Brazil has seen the emergence of national movements defending the interests of citizens affected by the construction of hydroelectric power stations. Such is the case of the Movement of Dam-Affected People (MAB). Grassroots movements like the MAB are a measure of how local people can get organized in order to resist the interests of external actors, and the close study of such phenomena showcases differences in social organization between affected populations in the south and in the north.

According to Sigaud (1995), families affected by dam construction in the south reacted much more quickly than families in the northeast, and the birth of the MAB exemplifies precisely that. In the mid-1980s, Eletrosul advanced an ambitious plan to build more than 20 hydroelectric plants around the Uruguay River. Local leaders were ready to react, seeking ways to inform the local population about what was going on. The MAB—formerly known as Regional Commission of Dam-Affected People—was created, primarily, to provide training and independent information to locals, functioning as a counterpoint to official communications provided by the Eletrosul and the federal government—or the lack thereof. As a result, local population was prepared for the arrival of the plants long before the construction begun. In the northeast, on the other hand, local populations displayed a feeling of "disbelief" in the face of events and an unreadiness to prepare and organize (Sigaud 1988). Although Sigaud's assessment has to do with events that took place in the 1980s and 1990s, it can be helpful in tackling the situation experienced by the Altamira population and the ways wherein families in the north and the northeast have reacted to similar circumstances.

In the case of Tucuruí, a group of dispossessed families got mobilized in September 1982 (Pinto 2012) to demand reparations and proceeded to create the Movimento de Expropriados de Tucuruí (Magalhães 2007). In the north, social movements had a greater importance denouncing damage done to the environment, a very relevant role in the specific context of the Amazon forest. Also, those social movements were very important in the process of rethinking the model of plant construction prevalent in the 1970s and the 1980s; as a result, dams like those of Tucuruí or Balbina are no longer built in Brazil. Other dams with vast environmental impact may be designed in the future, but it is very difficult that any of them have water supplies as big as Balbina's. When it comes to local populations themselves and their day-to-day life, social movements were less relevant in the north than they were in the south.

The shortcomings of northern social movements were apparent in their wavering struggle against the Belo Monte Dam. Their efforts were greatly harmed by the lack of organization and unity. The most active group currently protesting against the dam is the one created by indigenous tribes whose territories were impacted by the power plant. Ethnic groups like the Parakanã, the Curuanã, the Xipaya, and the Asuriní held several protests at the construction site or in front of the Norte Energia offices. According to the prosecutor of the Federal Public Ministry, Thais Santi, these actions bore fruit and several of their complaints were heard. One of the reasons for their success might be the special treatment afforded to indigenous people by the government and the company. Other movements might leave their own mark, but hold less bargaining power. One of them is Xingu Vivo para Sempre (Xingu Alive Forever), a movement that congregates "riverines, fishermen, workers, rural workers, indigenous, inhabitants of Altamira, people affected by dams, women's movements and religious and ecumenical organizations" (Xingu Vivo para Sempre). The MAB is also present in Altamira, but is much less strong there than it is in the south, for political and structural reasons. In February 2014, during an interview, a local leader has stated that "the MAB can't fight" against the company. In 2015, the Forum de Defesa de Altamira tried to bring together all different actors, but it never gained enough local influence to make its voice heard amid the political games.

All over Brazil, either in the north or in the south, the government has failed to provide proper and full compensations to families affected by dam construction. In the south, thanks to an agreement reached between Eletrosul and the Regional Commission of Dam-Affected People in 1987, families were offered different options of compensation: They could either choose to be paid money or to be separately or collectively resettled in urban or rural areas. However, the agreement between Eletrosul and CRAB has no legal value: It is merely a means for both sides to negotiate. Therefore, there is no legal provision establishing compensation options for the affected families.

In the north, according to an interview with Public Defender Andrea M. Barreto in February 2014, Norte Energia has not complied with the social obligations it assumed, especially toward the families in the Santo Antônio community. Most of those families ended up opting for money compensations, although their first choice had been rural relocation for the entire community. All over the rural area, as a matter of fact, local populations have abandoned the idea of being collectively resettled. According to studies by the Instituto Socioambiental (ISA) and the Getulio Vargas Foundation, Norte Energia has failed to fulfill commitments taken under the Environmental Impact Assessment and Environmental Basic Plan (PBA), which hinders family relocation. Families have staged several marches demanding the regularization of compensations, with the formal support of IBAMA.

The great difference between social movements in the north and the south lies in action and reaction dynamics. In the south, preemptive actions hindered the forceful migration of affected families, whereas in the north the social movements arose to claim better compensations for families that already had changed their abode. In both cases, "the conflict established by the implementation of hydropower projects exemplifies the struggle for environmental justice, revealing us the dispute over the social reappropriation of nature" (Xingu Vivo 2016).

3.2.2 *Differences in Spatial Organizations*

Analysis of the distribution and organization of families is the base to population studies of space. The main distribution of population, in both areas, in the north and in the south of Brazil, occurs linearly alongside the course of a river. Therefore, the same pattern can be observed in all verified areas: Major constructions impact first on families that live on the riverbanks. In both cases, there is a strong connection between riverine families and natural elements. This relationship is based on natural resources like water and fish and also on the leisure and beauty associated with the physical structure of the landscape. Despite the linear aspect, there are specific differences between the two regions based on other spatial structuring elements.

The temporality of the families and the social and territorial construction are two essential characteristics that must be tackled in order to understand the relationship between the families affected by the dam in the south and their territory. According to archaeological research, populations formed by "hunter-gatherer groups" (Onghero and Frasceschi 2009) left settlement marks on the Uruguay River dating back to 10,000 years ago. The river was the main source of livelihood to the indigenous groups that occupied the region, especially the Tupi-Guarani and Kaingang (Renk 2006). A significant increase in population would occur only in the early twentieth century (Bellani 2014). The west of Santa Catarina State and the northwest of Rio do Grande do Sul State were very remote areas for the first settlers; therefore, they were the last to be occupied by European colonization. This circumstance defined population patterns in both states. On the banks of the Uruguay River, the first European settlers were called "pathfinders," an evidence of how empty and isolated the region was at that time. The population concentrated along the river not only for timber exploitation from the banks, but also because that location made it easier for people to move toward Argentina. The countryside, far from the riverbanks, was occupied much later. Some regional urban centers, like Erechim (RS) and Chapecó (SC), would appear only in the twentieth century.

In 2007, while researching populations affected by the hydroelectric power plant of Foz do Chapecó (Leturcq 2010), we observed that population identified deeply with their living space. Sixty-five affected families were interviewed in six municipalities in the vicinities of Foz do Chapecó (in Rio Grande do Sul and Santa Catarina states). Of those families, 30 had come to their respective regions between 1960 and 1979. Half of the 65 householders were born in the same city of residence, and the other half was born in neighboring municipalities, with a maximum distance of 100 km. A total of 24 families had never left the city where they lived. The importance of the relationship between these families and the place they inhabit is evidenced by the prolonged time of residence, and the sense of belonging of individuals toward a particular region is strengthened by frequency and time.

In the north of the country, the population is nowadays much scarcer. After all, we are talking about the Amazon rainforest, whose original inhabitants were either "persecuted or decimated" and are now scattered through the forest. Even before the arrival of European settlers, the region was not densely populated: It harbored around 8 million people by the end of the fifteenth century (Clement et al. 2015), who lived in small villages or led a nomadic life. Indigenous population has decreased dramatically after contact with Europeans: The mortality of native inhabitants in the Amazon over the last centuries has been as high as 90%. Until the eighteenth century, they were the majority population; since then, they have become a minority group. Nowadays, only 2.1% of the population in the Amazon is recognized as indigenous, according to the 2010 IBGE census; in Brazil as a whole, they represent only 0.4% of the population. From 2000 to 2010, on the other hand, indigenous population has increased in 11%, mainly for identity reasons, a phenomenon described by IBGE, as "ethnogenesis or ethnicization" (IBGE 2010). In the twentieth century, that population has passed from a series of large-scale complex societies to small-scale societies (Clement et al. 2015). This is the context wherein power plants began to be created on the Amazon, a region where indigenous populations had been losing influence over the centuries. Most people living on the Amazon nowadays are the result of recent migrations that followed development cycles, such as the rubber cycle (early twentieth century) or the building of the Trans-Amazonian Highway (during the 1970s).

In terms of socio-territorial construction, rural populations of European descent in Brazil have organized themselves around the church in the form of community collectivities, as described by Henri Mendras: "The peasant societies (…) are organized into collectivities, small and relatively autonomous, installed on a territory they exploit" (Mendras 1995). In the municipalities established along the river Uruguay by German and Italian immigrants, we find communities centered around a church; their transport routes are through dirt roads or rivers. Such riverain social structures create family connections and neighborhood interrelationships.

Spatial distribution is much different in Altamira, Lower Xingu, in the north of the country. The occupation of territory in this region has many specific traits that set it apart from the rest of Brazil. The area has been colonized during the

Trans-Amazonian Highway construction, from 1970 to 1990. Migrants from the south and the northeast moved to the region in order to colonize lots of land offered by INCRA (Moran 1975) alongside the burgeoning road.

During the settling of the area by migrants from 1970 to 1990, the occupation of the land followed two main patterns. Some families decided to live in their own lots, while others constituted rural villages, where families and activities were concentrated. Therefore, several population centers can be found in the construction zone of Belo Monte. Agrovila, for instance, lies at the 60 km, while Santo Antonio Center is located near the side road 50. In addition to that, small farmers can be found scattered in separate lots. A distinct and significant population present in the Altamira region are the fishermen and small coastal farmers who live on the banks of the Xingu River. They live scattered and isolated along the banks, showing a strong connection (both in economic, social cultural, and commercial terms) with the city of Altamira, which they visit very often. Field observations confirm that the rural–urban connection is strong, and people used to live in the two spaces. The only nearby rural communities on the Xingu River are located on islands, as the Ilha da Fazenda, in Volta Grande, about 40 km from the city. According to the Environmental Impact Assessment EIA-2009, there are approximately 350 riverine families established on the island, which will be directly affected by the Belo Monte Dam (Eletrobrás 2009).

Interestingly, the increase in indigenous populations in Altamira between 2000 and 2010 was even higher than in the rest of the country, reaching 21.8% (IBGE 2010). To explain this growth, we follow the same criteria as the IBGE: Nowadays, more people identify as "indigenous" than they did in previous decades. Even so, indigenous population represented slightly less than 4% of all the Altamira population in 2010.

It should be noted that the company responsible for the Belo Monte power plant provided indigenous peoples with a differentiated treatment, therefore promoting the indigenous self-identification. Numerous riverine families in the region decided to "become" indigenous to have access to specific benefits, no longer being in a "invisible" affected riverine category (following the interview with the prosecutor of the Federal Public Ministry, Thais Santi in Altamira, May 2015). In the south, building contractors gave no special treatment to indigenous populations during the dam construction. This happens because indigenous populations hold less influence in the south than in the north. Therefore, when an indigenous group is affected by a project, as happened in Foz do Chapecó and Monjolinho, they do not get the same kind of special treatment afforded to Indians in Belo Monte.

Construction impact on urban households is a socio-territorial element insufficently addressed by specialized literature and technical analysis. Altamira, affected by the Belo Monte power plant, and Itá, affected by the Itá power plant in Santa Catarina, are two examples of urban sites that have been directly impacted. In 1996, Itá, a city with 2000 inhabitants (of a total of 8000 people in the entire district) had to be rebuilt on a higher altitude, as the old town was to be flooded by the waters of the power plant. The new city, planned for 2500 inhabitants according to the 1995 Eletrosul information, had more than 4000 people in 2010 (IBGE 2010). The case of Itá is extreme, because the moving of the entire town to a different place costed a lot of money and represented a very hard job for Eletrosul, the company in charge. According to Raquel de Mattos Viana:

> The case of Itá is emblematic because, despite a vast initial investment in the relocation, what we see nowadays, after the construction of the plant, is a series of economical problems and high levels of unemployment. The urban question, in the case of Itá, appears first and foremost as an urbanistic problem (Viana 2003).

Altamira is a different case because the city was not relocated or reconstructed in its entirety. Altamira had 76,700 inhabitants in 2010 (IBGE 2010), and therefore, the number of people affected was much higher there than it was in the south. In Altamira, according to Norte Energia, 4,362 families were affected—a total of 16,420 people (RIMA 2009). After the migration of the urban families, no further information is available. What we know is that 3,587 families were relocated in collective rural settlement on the outskirts of Altamira (Norte Energia 2017). We do not know the background of those families, but it is known that they came both from rural and urban areas. Another impact of the Belo Monte power plant has to do with Altamira's urban structure. An average-sized city in Brazilian terms, Altamira suffered several changes in its urban configuration. The first structural change occurs at the wharves on the Xingu River. The Norte Energia, in its Basic Environmental Project, says that "there should be deployed all necessary recreation infrastructure, including toilets, kiosks with grills, sport spaces and docks for recreational boats" (Onghero and Franceschi 2008). As we observe, the city today, such improvements are not to be found, except for the building of an artificial beach. Another element effected a profound change: the main urban stream of the Xingu, which cuts the city in half. The urban part of Xingu River will be larger both in height and in width, with a new area of permanent protection of 100 meters (Norte Energia 2011). In order to establish a comparison, we can include the impact on the urban population of the Madeira River dams, also in the north. The Santo Antônio power plant is located 7 kilometers from the center of Porto Velho, home to more than 250,000 people (IBGE 2010). The total number of people affected in the urban area is 1100 (Cavalcante and Santos 2012), but the urban impacts are not highlighted in the process of impact reviewing. In the case of Tucuruí, Eletronorte estimates that more than 1,400 urban families (about 7,000 people) were affected or had to be relocated.

3.2.3 *Results, the Redistribution of Families*

The building of power stations always generates spatial and territorial changes: When a population is directly affected, some sort of migration ensues, and a new life must begin within a new territory. Although the building of a dam always causes similar situations in different parts of the country, great differences were observed between the south and the north. Those differences are especially noticeable when one examines what happened to the affected families after they arrived in their new place of residence. Differences arise due to the previous history of each population and its distribution patterns. Socio-territorial specificities shape different reactions to a similar situation. In the south, families have been struggling to rebuild their lives for almost 30 years. In the north, the struggle has just begun and the process of redistribution is still taking place—which makes it difficult for a full comparison to be made. In both cases, the consequences have a direct impact on the lifestyle and the living space of the affected families.

In the south, where five large dams were built along the Uruguay basin, most compensations were paid in money. It is not clear if the affected families opted freely for the money compensation or if they were forced to accept it by a lack of better alternatives. Be it as it may, money compensation is clearly the option that offers more freedom of choice both for the family and the company. Among those who did not take the money compensation, most opted for the letter of credit, which enabled them to move individually to a new property. Only 9% of the affected families opted for collective rural relocation—which is curious, because this modality offers the best conditions for building a new life. In addition to receiving new lands and new houses, families that opted for this kind of compensation were resettled in communities with new roads, electricity, schools, etc. They were also granted social and technical assistance for 5 years. One might argue that people who opted for money compensations made their choice without a full understanding of the situation.

In Altamira, most compensations are taking place in urban areas. Norte Energia begun construction of six Urban Collective Resettlements (RUC), each now in a different state of construction, spread through 316.5 ha around the city (Cf. Fig. 2.5). The Jatobá RUC, for instance, now houses hundreds of families—although it remains partly unfinished. Some of the families who were given urban residences had a rural background. According to prosecutor Thais Santi, this happens because fishermen and riverine populations in general are used to have two places of residence, one in town and other closer to the river. At the time of the resettlement, families were made to choose an official residence, and many opted for a house in the RUC. Therefore, they had to adapt to a new spatial reality: a urban space with new neighbors and different means of mobility. The prosecutor also confirmed that Native American families have being temporarily settled in urban houses.

It is not possible to state with certainty that those families were force to move to the city. However, it is clear that the affected population had few choices: If they did not take money or a house in Altamira, they would have to wait for collective rural relocation or for a new settlement to be built near the Xingu River. Norte Energia is presently building a new settlement named Petral, occupying 500 lots on the southern edge of town, close to the riverbanks. This place will be occupied both by fishermen and Native American families.

References

Bellani EM (2014) Balsas e balseiros no Rio Uruguai (1930–1950). Revista Cadernos do Ceom., Chapecó 19(23):73–98

Bonin S (2004) Au-delà de la représentation, le paysage. Strates, 11. http://journals.openedition.org/strates/390

Cavalcante MM, Santos LJ (2012) Hidrelétricas no Rio Madeira-RO: tensões sobre o uso do território e dos recursos naturais na Amazônia. Confins 23(15)

Claval P (1972) La réflexion théorique en géographie et les méthodes d'analyse. L'espace Géographique 1:7–22

Clement CR, Denevan WM, Heckenberger MJ, Junqueira AB, Neves EG, Teixeira WG, Woods WI (2015) The domestication of Amazonia before European conquest. Proc R Soc B 282:20150813

Eletrobrás, Eletronorte, MME (2009) Relatório de Impacto Ambiental—RIMA. Aproveitamento Hidrelétrico de Belo Monte. http://eletrobras.com/pt/AreasdeAtuacao/geracao/belo_monte/Rima-AHE-Belo-Monte.pdf

Fearnside PM (2017) Planned disinformation: The example of the Belo Monte Dam as a source of greenhouse gases. In: Issberner LR, Lena P (eds) Brazil in the anthropocene: conflicts between predatory development and environmental policies. Taylor & Francis Group, New York, Routledge, p 368

IBGE—Instituto Brasileiro de Geografia e Estatística (2010) Censo Demográfico 2010. IBGE, Rio de Janeiro

Leme Eletrobrás (2009) Aproveitamento Hidrelétrico (AHE) Belo Monte—Estudo de Impacto Ambiental (EIA). Eletrobrás, Rio de Janeiro

Leturcq G (2010) Migrations forcées dans le Sud du Brésil: les *atingidos*. PhD Thesis, Université du Maine, Le Mans

Magalhães SB (2007) Lamento e dor. Uma análise sócio antropológica do deslocamento compulsório provocado pela construção de barragens. PhD Thesis, Universidade Federal do Pará and Université Paris, Belém

Mendras H (1995) Les sociétés paysannes. Gallimard, Paris, p 368

MME—Ministério de Minas e Energia (2007) Plano Nacional de Energia 2030. Ministério de Minas e Energia and Empresa de Pesquisa Energética, Brasília

Moran EF (1975) The Brazilian colonization experience in the transamazon highway. Anthropology 16(1):29–57

Movimento Xingu Vivo para Sempre (2016) Quem Somos. MXVPS. http://www.xinguvivo.org.br/quem-somos/

NE—Norte Energia (2011) Projeto Básico Ambiental das instalações iniciais do aproveitamento hidrelétrico Belo Monte

NE—Norte Energia (2017) 12 Relatório Semestrais UHE Belo Monte

Onghero AL, France LA (2008) Vidas em torno do rio: uma abordagem histórica das relações estabelecidas entre as populações na região Oeste de Santa Catarina e Noroeste do Rio Grande do Sul e o Rio Uruguai. Revista Cadernos do Ceom, Chapecó 21(29):107–132

Pinto LC (2012) Os projectos hidrelétricos como causa dos deslocamentos populacionais: migrações forçadas em nome do desenvolvimento. Master Dissertation, Universidade Nova de Lisboa, Lisboa

Renk A (2006) A luta da erva: um ofício étnico da nação brasileira no oeste catarinense. Chapecó, Argos, p 250

Reynaud A (1984) L'intérêt de la démarche comparative en géographie. Espaces Temps 26-27-28:26–33

Rodrigues JN (2012) La comparaison en géographie. Contribution à partir de l'étude de la coopération intercommunale en France et au Brésil. Confins 16. http://journals.openedition.org/confins/8062

Santo Antônio Energia (2015) Mobilidade Urbana trará melhorias a portovelhenses. http://www.santoantonioenergia.com.br/mobilidade-urbana-trara-melhorias-a-portovelhenses/
Sigaud L (1988) Efeitos sociais de grandes projetos hidrelétricos: as barragens de Sobradinho e Machadinho. Rosa LP Impactos de grandes projetos hidrelétricos e nucleares: aspectos econômicos, tecnologicos, ambientais e sociais. Marco Zero, São Paulo, pp 83–165
Sigaud L (1995) Estado, energia elétrica e meio ambiente: o caso das grandes barragens. COPPE/Editora da UFRJ, Rio de Janeiro, p 184
Souza AA (2011) Itaipu e a urbanização da zona de fronteira do Iguaçu: cidade e conjuntos habitacionais da usina hidrelétrica. Master Dissertation, PUC-Campinas
Verdum R (2011) Os geógrafos frente às dinâmicas sócio-ambientais no Brasil. Revista do Departamento de Geografia, São Paulo 16:91–94
Viana MR (2003) Grandes barragens, impactos e reparações: um estudo de caso sobre a barragem de Itá. Master Dissertation, Universidade Federal do Rio de Janeiro, Rio de Janeiro

Conclusion

Hydroelectric dams are an integral part of the Brazilian history, not only for their production of energy but also for their contribution to the national landscape. They accompany the major political and economic phases of the country and mark the various historical periods. These iconic dams symbolize great moments of the country and the evolution of society. The social and territorial markings of dams are ongoing in the history of electric power in Brazil, and they evolve with the times, between periods of progression and regression. Their imprints may be more or less significant but never cease to exist.

The construction of dams and their various associated consequences raise questions on the development models adopted by successive governments, and for decades, and in all cases, the population has been kept out of the decision-making process.

Although governments and economic situations change over time, the way in which dam-affected families are treated does not fundamentally change. Evolutions nevertheless still exist, both positive and negative. For instance, family protection rights seem to protect them faster but the loss of state influence leaves the door open to economic entities who have the final say in the decisions.

The social acceptance of dams varies depending on the families and the situation, but it is important to note in this conclusion the discordance between what was presumed and the actual reality. Some dam-affected families in Brazil accept and understand the construction of the structures and their consequences. They are aware of the environmental repercussions but keep a positive view of the work on different levels. Figure A1 shows the opinions of more than 260 families affected by the Belo Monte Dam on the effects of the dam at various scales. Forced to migrate, these families received compensation for their losses and resettled in the new urban district of Jatobá, which is not without structural problems.

Overall, we see a positive opinion of dams, particularly at the national level (70% positive), thus demonstrating the effects of advertising produced by the federal government and the company presenting the structure as essential to the country's future. The only area where the influence of the dam is mainly described

G. Leturcq, *Dams in Brazil*, SpringerBriefs in Latin American Studies,
https://doi.org/10.1007/978-3-319-94628-3

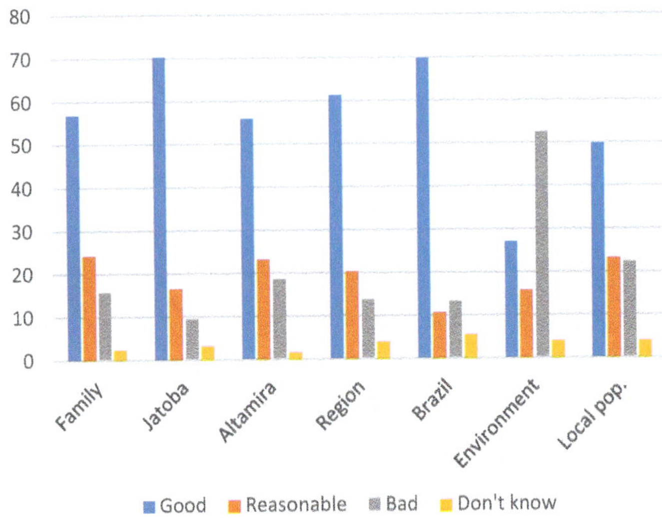

Fig. A1 Acceptation of Belo Monte by affected people

as negative is the environment (52.5%). Families are aware of the impacts of the construction on the natural environment, especially in the Amazon. The contrast is interesting between the social difficulties experienced by the families and the acceptance of the dam, triggering factor of the disturbances.

The various problems associated with the construction of a dam could be solved by a significant increase in the involvement of the local populations in the process, with a potential revision of the implementation schedules. This would allow a real participation and a possibility of exchanges between the government, the people, and the dam company. The lack of space for real dialogue is evident, particularly because of the counterproductive policies put in place by the companies to invalidate the relationship with the population in order to enjoy complete freedom of action.

The arrival of a dam and its repercussions affect not only the people but also the whole society. Local families are the most affected. There are many ways of reducing the consequences of a dam construction, and decision makers are well aware of the numerous solutions, plans, and strategies. What is missing is a political and economic will, notably on the part of the dam company, to have a lesser social impact. It is therefore necessary to empower the stakeholders to mitigate the impacts. There is no shortage of potential solutions (time and money in particular), but their implementation is very weak, and the little effort made by companies clearly indicates their lack of involvement and social responsibility.

Dams and the companies responsible for them are therefore factors of increasing social, territorial, and environmental inequalities. The construction of a dam affects poor families in the region even more than the middle classes who benefit from resources (economic, political, cultural, information, etc.) to take advantage of the opportunity. Vulnerable families, such as the southern caboclos or the northern ribeirinhos, are the people who suffer the most from dams as they lose the little they had and gradually need to rebuild their lives in a new environment sometimes in precarious conditions. The dam construction process therefore causes territorial inequalities.

Mitigating measures could be taken, but politicians and businesses would need to want it. For example, an electricity bill exemption for expelled populations seems like a sensible option and would demonstrate an attempt to ease conflict and to reduce social inequalities.

Population migration is the main consequence of a dam construction. It influences the families' integration into the new living space and changes the degree of future mobility. Families who fail to settle quickly in the new environment look for a new place of work. This requires greater mobility, which is enabled by the recent acquisition of new modes of transportation, also allowing more visits to relatives and friends from the old communities.

The concept of proximity is essential to understand the new mobility induced by dams. Most families agree to migrate knowing the inevitability of the phenomenon, but, in addition to their lack of participation in the compensatory process, they have no say in the location of their final destination despite wanting to stay as close as possible to their old place of life. The living space and the families' migratory history are fundamental elements to minimize the effects of dams. The more families are used to moving and adapting to new compatible living environments, the more chance they will have to successfully settle into their new living space. Thus, it is recommended to provide social support to forced migrants who have deep roots in their territory and no migratory experience, as opposed to regular migrants who are already aware of the risks and consequences of displacement. The other dimension analyzed in the adaptation of local populations to the construction of a dam concerns the preparation of the affected families. While the means of communication were more limited in the 1980s, the populations of the Uruguay River seemed better prepared for the arrival of the dams. Today, dam-affected families, who had high expectations, are surprised by the negative effects that arise toward the end of the project. One can therefore wonder about the preparation and the circulation of information regarding the tangible and recurring impacts of dams.

People's opposition to the economic model and its negative consequences seems to diminish over time in the face of an imposing dominant system that leaves no room for negotiations. While the population was considered of no consequence during the military dictatorship, the activists were able to organize themselves and make their voices heard to claim and get an opportunity for constructive discussions in the 1980s and 1990s. Today, this opportunity has greatly diminished. The representatives of dam-affected families are less listened to and companies much prefer to deal with individual cases. MAB seems less influential than at the turn of the twenty-first century, allowing us to question the strategies put in place by the movement and its future place in society and the electricity sector. Pockets of resistance still remain, like the local leaders of Itapiranga who have been fighting for 30 years against the construction of a dam in western Santa Catarina. However, they still suffer the consequences (early departures in particular) and remain at the mercy of a change of local government, which could thwart their efforts, as is currently the case.

The relationship between the private and the public sectors is also a central issue raised in this study where it can be seen that connections, somewhat official, often undermine the common good and serve above all the interests of the companies. Corruption, one of Brazil's ills, is not exempt from the construction of hydroelectric plants, as shown by the suspicion and denouncement of a 150 million real payment during the construction of the Belo Monte Dam. Current practices need to be regulated in Brazil so that the public interest is back at the center of politics, local as well as national and regional, but this effort should not be limited to the electricity sector.

The construction of hydroelectric plants will continue in the medium-term future of the country, particularly because of the continuity of the national energy policy and because of the slow changes of direction that the country is trying to take. The territorial effects of dams are known to the developers, and to minimize the consequences, some future dams, such as the Garabi energy complex, are trying to join a process of international integration with neighboring countries. The national political will go in this direction, and only the economic and political contexts will dictate the next developments in terms of dam construction.

The impacts of dams in Brazil have long been researched. There are some very comprehensive studies focusing on a wide range of topics and disciplines. However, we still notice some shortcomings especially with regard to migration and mobility. The effects of proximity are obvious but appear as a first stratum of analysis. The wide scope of the subject allows us to think of potential future research, especially with regard to economic migrants: opportunity seekers, dam workers, rural exodus, etc. We can also envisage research projects on the link between migration history, forced migration, and adaptation to the new place of life. The correlation between the three phenomena presents a vast array of problems, and the examples encountered during our research suggest that a relationship does indeed exist. We can raise many questions, especially in view of improving the attention given to migrant families with the aim of facilitating their resettlement in the new living spaces. The construction of dams in Brazil appears inevitable for the next few decades, but it is now essential to try to change the balance of power so that the most fragile local populations do not pay the price of energy policies.

Index

G. Leturcq, *Dams in Brazil*, SpringerBriefs in Latin American Studies,
https://doi.org/10.1007/978-3-319-94628-3

The manufacturer's authorised representative in the EU is Springer Nature Customer Service Centre GmbH, Europaplatz 3, 69115 Heidelberg, Germany. If you have any concerns regarding our products, please contact ProductSafety@springernature.com

Printed and bound by CPI Group (UK) Ltd, Croydon, CR0 4YY
22/09/2026
02220957-0010